User's Manual for the Graphical Constituent Loading Analysis System (GCLAS)

By G.F. Koltun, Michael Eberle, J.R. Gray, and G.D. Glysson

Techniques and Methods 4-C1

U.S. Department of the Interior
U.S. Geological Survey

U.S. Department of the Interior
DIRK KEMPTHORNE, Secretary

U.S. Geological Survey
P. Patrick Leahy, Acting Director

U.S. Geological Survey, Reston, Virginia: 2006

For product and ordering information:
World Wide Web: http://www.usgs.gov/
Telephone: 1-888-ASK-USGS

For more information on the USGS—the Federal source for science about the Earth,
its natural and living resources, natural hazards, and the environment:
World Wide Web: http://www.usgs.gov
Telephone: 1-888-ASK-USGS

Suggested citation:
Koltun, G.F., Eberle, Michael, Gray, J.R., and Glysson, G.D, 2006, User's manual for the Graphical Constituent Loading
Analysis System (GCLAS): U.S. Geological Survey Techniques and Methods, book 4, chap. C1, 51 p.

Contents

Figures

Tables

Conversion Factors

Multiply	By	To obtain
Length		
foot (ft)	0.3048	meter (m)
Volume		
gallon (gal)	0.003785	cubic meter (m^3)
million gallons (Mgal)	3,785	cubic meter (m^3)
cubic inch (in^3)	16.39	cubic centimeter (cm^3)
cubic foot (ft^3)	0.02832	cubic meter (m^3)
acre-foot (acre-ft)	1,233	cubic meter (m^3)
Flow rate		
cubic foot per second (ft^3/s)	0.02832	cubic meter per second (m^3/s)
million gallons per day (Mgal/d)	0.04381	cubic meter per second (m^3/s)
Mass		
pound, avoirdupois (lb)	0.4536	kilogram (kg)
ton, short (2,000 lb)	0.9072	megagram (Mg)
ton, long (2,240 lb)	1.016	megagram (Mg)
ton per day (ton/d)	0.9072	metric ton per day (ton/d)
ton per day (ton/d)	0.9072	megagram per day (Mg/d)
ton per year (ton/yr)	0.9072	megagram per year (Mg/yr)
ton per year (ton/yr)	0.9072	metric ton per year (ton/yr)

Temperature in degrees Celsius (°C) may be converted to degrees Fahrenheit (°F) as follows:

°F = (1.8 x °C) + 32

Temperature in degrees Fahrenheit (°F) may be converted to degrees Celsius (°C) as follows:

°C = (°F - 32) / 1.8

Concentrations of chemical constituents in water are given either in milligrams per liter (mg/L) or micrograms per liter (µg/L).

Preface

This report describes the Graphical Constituent Loading Analysis System (GCLAS), a program developed by the U. S. Geological Survey (USGS) to facilitate computation of loads and average concentrations of physical and chemical constituents transported in streams. The documentation and program, including sample datasets, are available for download over the Internet from a USGS software Web page at **http://water.usgs.gov/software/surface_water.html**. Any future revisions or updates to GCLAS will be made available at the same location.

Although we have attempted to make GCLAS accurate and error free, complex programs such as GCLAS almost certainly will contain some errors. Users are encouraged to report any errors in this user's guide or in GCLAS itself to the contact listed on software distribution Web page.

User's Manual for the Graphical Constituent Loading Analysis System (GCLAS)

By G.F. Koltun, Michael Eberle, J.R. Gray, and G.D. Glysson

Abstract

This manual describes the Graphical Constituent Loading Analysis System (GCLAS), an interactive cross-platform program for computing the mass (load) and average concentration of a constituent that is transported in stream water over a period of time. GCLAS computes loads as a function of an equal-interval streamflow time series and an equal- or unequal-interval time series of constituent concentrations. The constituent-concentration time series may be composed of measured concentrations or a combination of measured and estimated concentrations. GCLAS is not intended for use in situations where concentration data (or an appropriate surrogate) are collected infrequently or where an appreciable amount of the concentration values are censored.

It is assumed that the constituent-concentration time series used by GCLAS adequately represents the true time-varying concentration. Commonly, measured constituent concentrations are collected at a frequency that is less than ideal (from a load-computation standpoint), so estimated concentrations must be inserted in the time series to better approximate the expected chemograph. GCLAS provides tools to facilitate estimation and entry of instantaneous concentrations for that purpose.

Water-quality samples collected for load computation frequently are collected in a single vertical or at single point in a stream cross section. Several factors, some of which may vary as a function of time and (or) streamflow, can affect whether the sample concentrations are representative of the mean concentration in the cross section. GCLAS provides tools to aid the analyst in assessing whether concentrations in samples collected in a single vertical or at single point in a stream cross section exhibit systematic bias with respect to the mean concentrations. In cases where bias is evident, the analyst can construct coefficient relations in GCLAS to reduce or eliminate the observed bias.

GCLAS can export load and concentration data in formats suitable for entry into the U.S. Geological Survey's National Water Information System. GCLAS can also import and export data in formats that are compatible with various commonly used spreadsheet and statistics programs.

Introduction

The Graphical Constituent Loading Analysis System (GCLAS) is a program designed by the U.S. Geological Survey (USGS) to compute daily loads of sediment or other stream-water constituents from time series of streamflow and concentration data. GCLAS computes loads as a function of an equal-interval streamflow time series and an equal- or unequal-interval time series of constituent concentrations. GCLAS was created to replace and improve upon the USGS Sedcalc program (Koltun and others, 1994), which was abandoned because software and hardware for which it had been developed became obsolete.

Primary features of GCLAS are the following:

- **A visual approach to data analysis.** GCLAS simultaneously displays data in graphical and tabular formats. Rather than typing in command lines to do computations and then generating plots to see the results, the analyst can manipulate graphical elements directly with a mouse or by editing data tables. Because the graphs and tables are dynamically linked, results of data manipulations can be seen immediately in both graphs and tables.

- **A comprehensive set of analytical tools.** In addition to providing editable graphs and tables already mentioned, GCLAS can

    1. display transport curves showing the relation between streamflow and concentration data being worked, a "reality check" for adding or repositioning estimated data points on a concentration curve,

    2. compute and (or) apply cross-section coefficients as a constant, as a function of streamflow, as a function of time, or combinations thereof, and

    3. compute loads for an entire year or any part of a year and export the load data to files for subsequent entry into the USGS National Water Information System (NWIS) (U.S. Geological Survey, 2003) or import into other analytical software.

- **The ability to compute loads for streams where reverse and (or) zero flows occur.** Because GCLAS makes it easy to resize, rescale, and zoom in within graphs, logarithmic transformations are not needed to constrain the y-axis, so working with zero and reverse flows is feasible.

- **Calculation of loads for constituents other than sediment.** GCLAS can compute loads for many constituents and can report loads in a variety of units.

Installing GCLAS

GCLAS was developed using the Java programming language, so it can be installed and run on computers with a variety of operating systems[1]. Installing GCLAS requires that a Java run-time engine (version 1.4 or higher) also be installed on the destination computer. It is recommended that GCLAS be installed on a system with at least 128 megabytes of RAM, a graphics card and monitor capable of displaying 1024 by 768 pixels or higher, and at least a 16-bit color display.

With the Java run-time engine installed, the GCLAS installer (which can be obtained from the USGS Water Resources Applications Software Web site at http://water.usgs.gov/software/surface_water.html) can be run by typing **java -jar gclas_install_1.05e.jar**[2] in a DOS or Unix window from the directory in which the jar file resides. Alternatively, with some operating systems (such as Windows) the GCLAS installer can be run by double left-clicking on the installer jar file. Proper installation requires that the person doing the installation have sufficient rights to create and write to directories where the GCLAS program and data will be stored.

After the installer has been started, the following windows will appear. (Note: The installation shown is for a computer with the Windows operating system. A Unix installation will result in a similar sequence; however, the installation paths will be different, and no option will be given to create a program group.)

---

[1]Although GCLAS should run on a variety of operating systems, testing of GCLAS has been predominately limited to the Microsoft Windows and Sun Solaris operating systems.

[2]As of this writing (May 2006), the current version of GCLAS is 1.05e. As new versions become available, the GCLAS installer's name will change to reflect the new version.

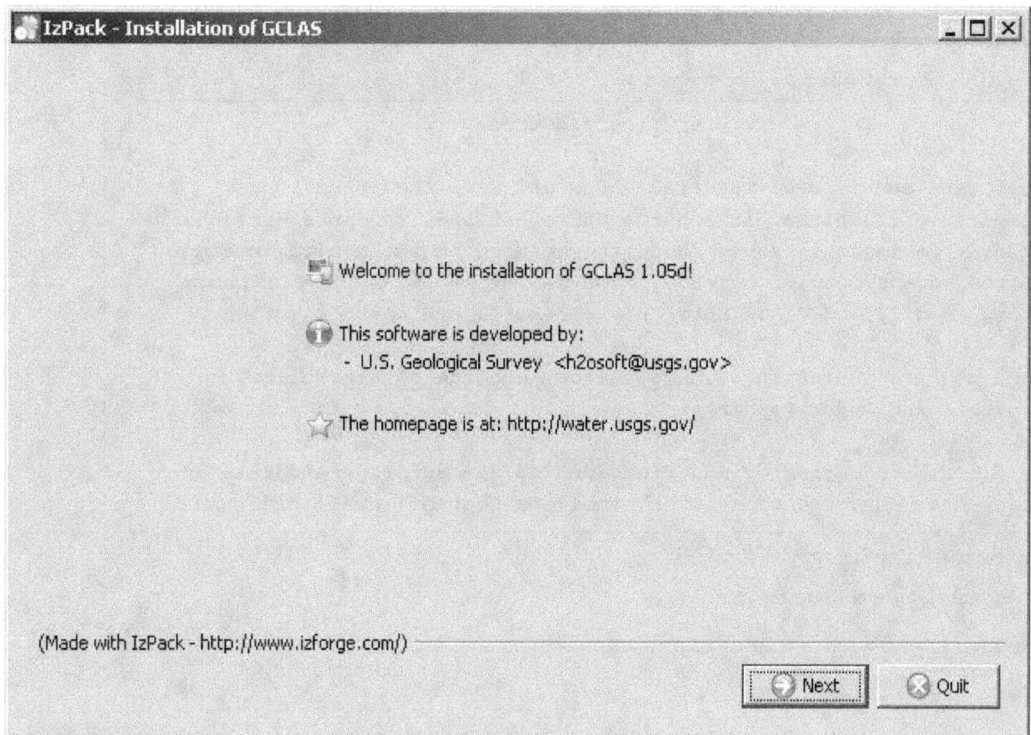

Click the **Next** button.

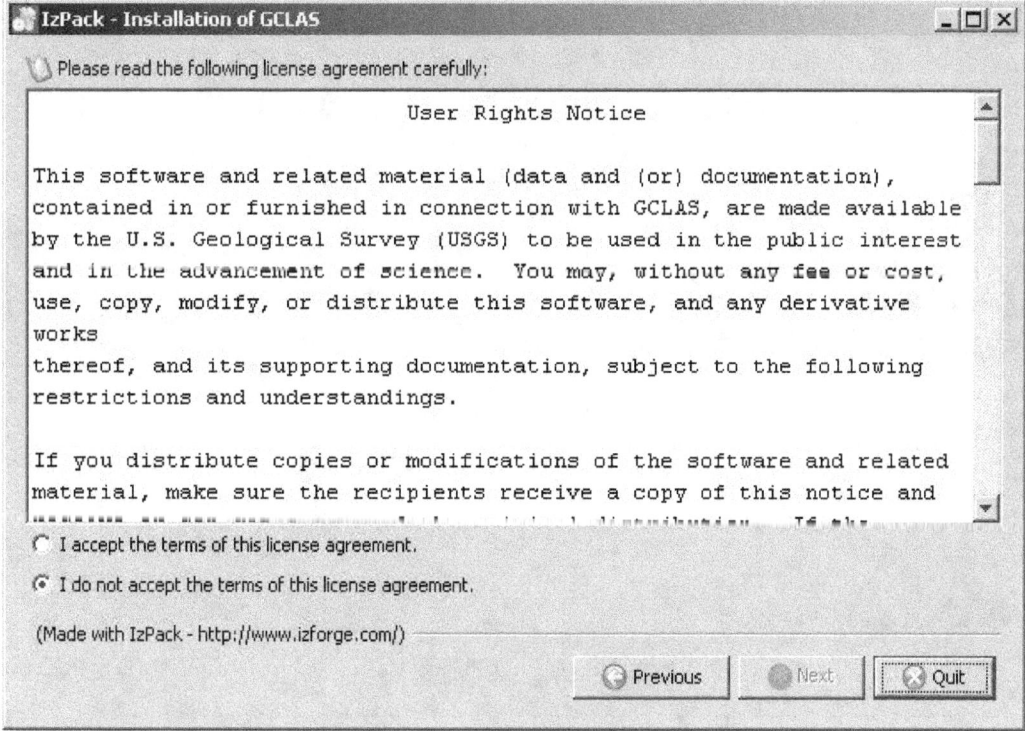

Click on the radio button to accept the terms of the license agreement.

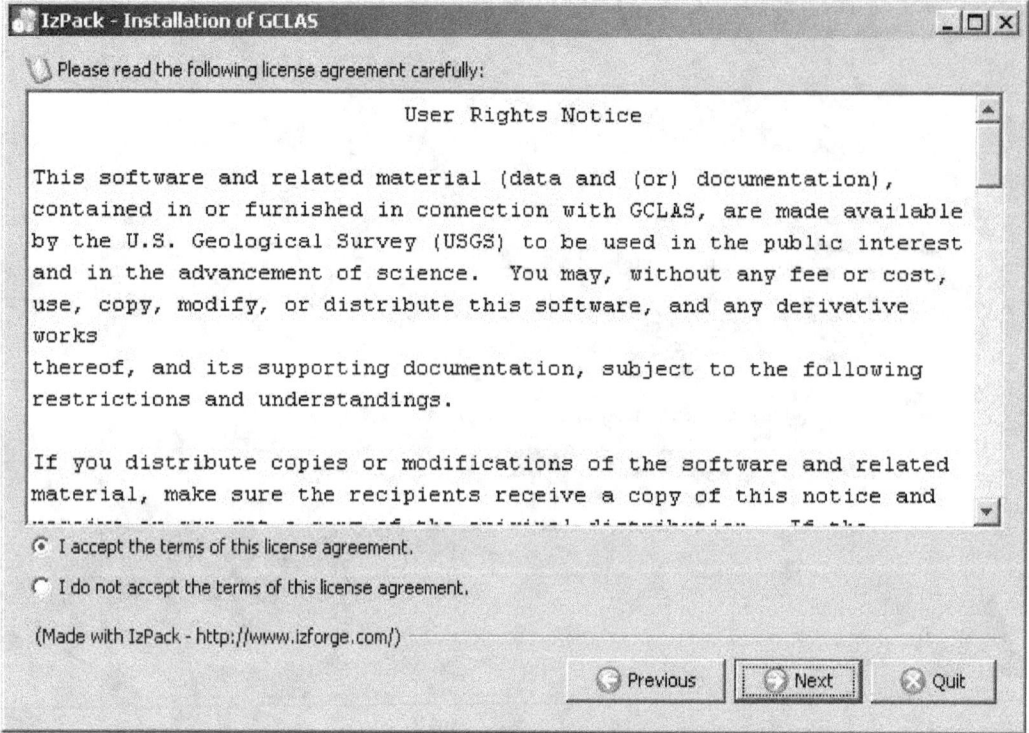

Click the **Next** button.

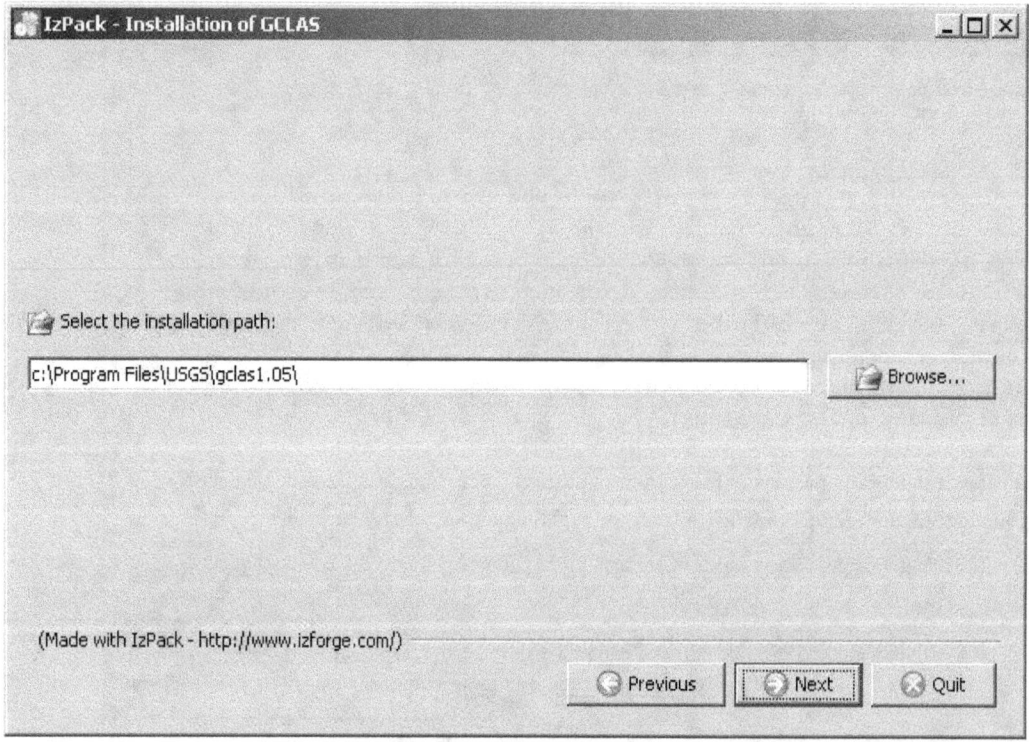

Accept or change the default installation path, then click the **Next** button.

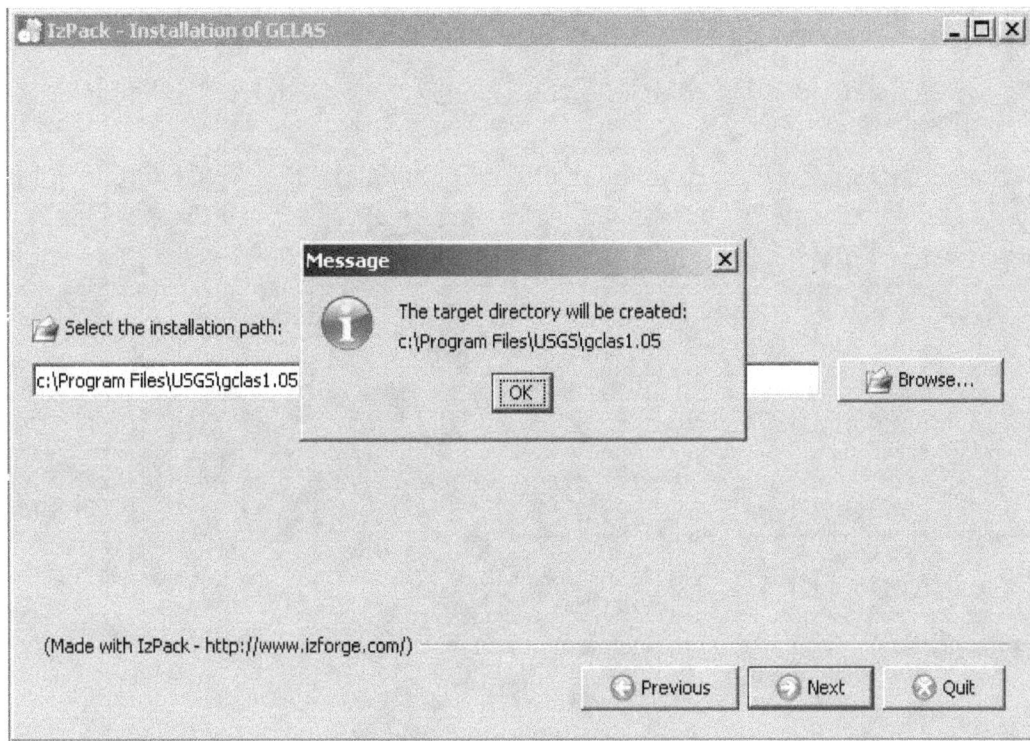

If the installation path does not exist, a pop-up message will inform you that the directory will be created. Click the **OK** button to proceed.

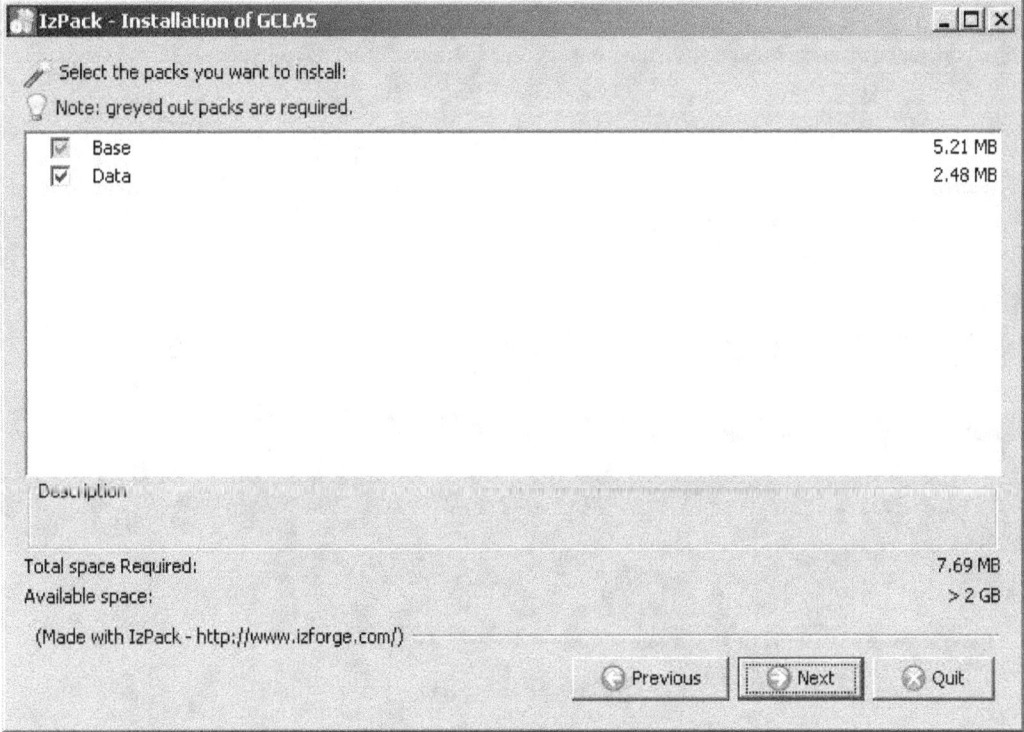

Choose whether you want to install the optional data pack (by checking or unchecking the adjacent box), then click the **Next** button. The data pack contains data input templates and sample input data sets.

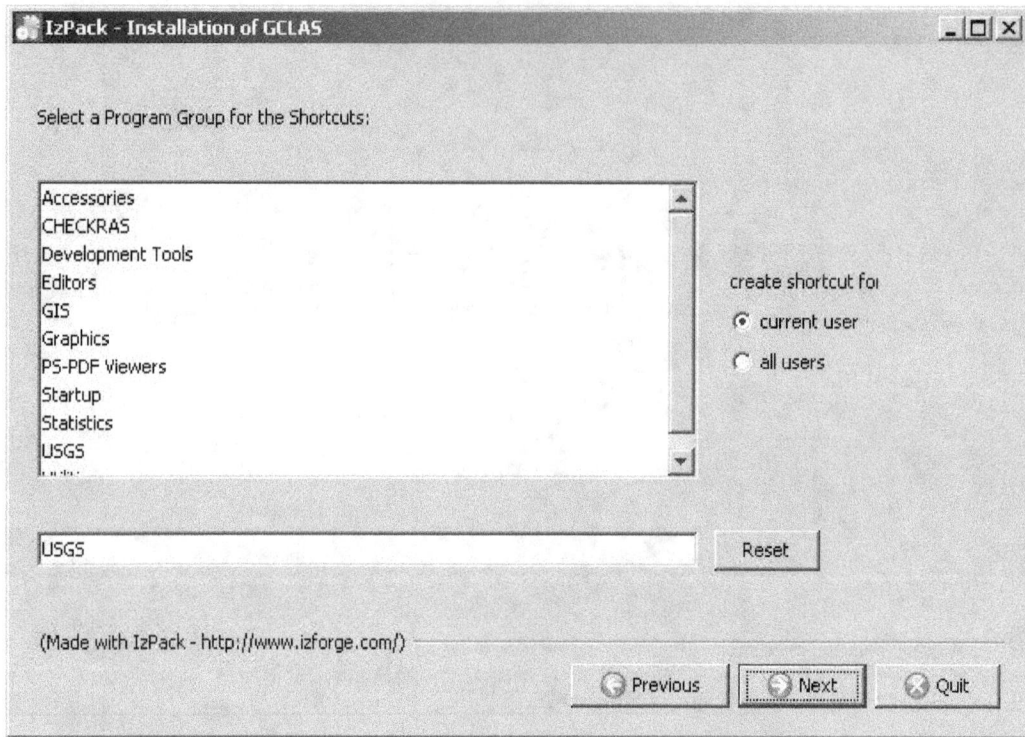

Accept the default menu program group for the shortcut (recommended) or supply a different program group, then click the **Next** button. Shortcuts can be created for the current user (the user installing GCLAS) or for all users (if the user has sufficient rights to do so).

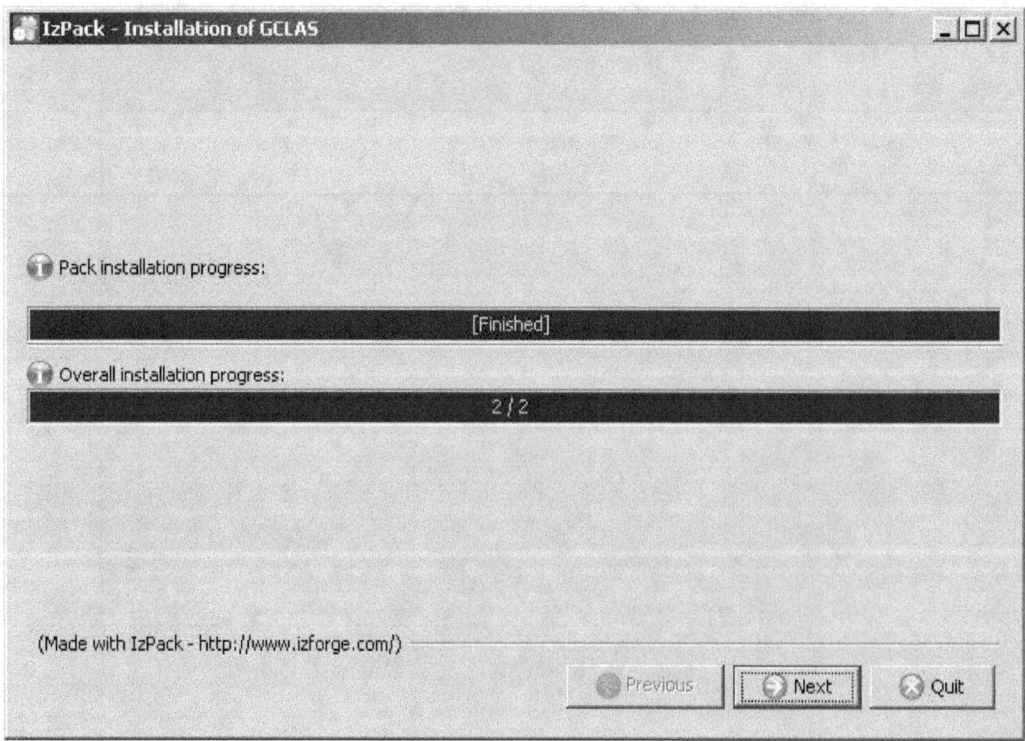

The installation progress will update and a window similar to that above will be displayed. Click the **Next** button.

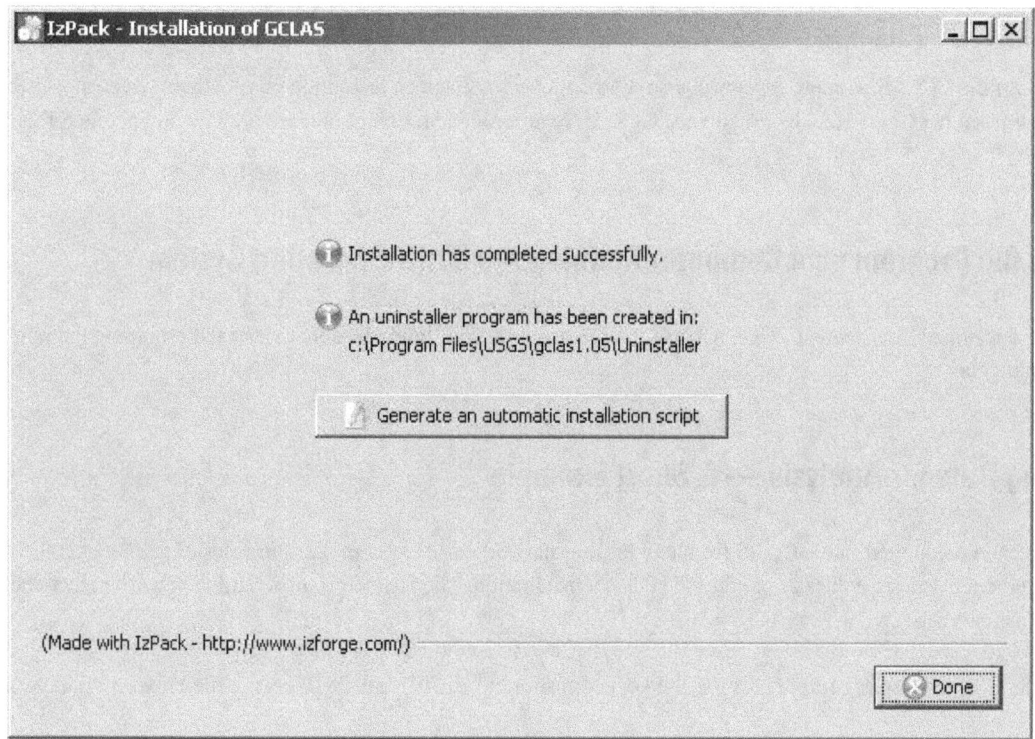

If the installation was successful, a window similar to that above will be displayed. Click the **Done** button to close the window. Optionally, an XML install script can be created by left-clicking the **Generate an automatic installation script** button. If, for example, the name autoinstall.xml is given for the install script, it can be copied into the same directory as the GCLAS installer on another machine, and then an identical install can be done without having to step through the prompt windows by issuing the command **java -jar gclas_install_1.05e.jar autoinstall.xml**.

If installation is done on a computer running the Unix operating system, a symbolic link to the gclas file in the bin directory of the GCLAS installation should be established in the user's path. For more information, see the unix_readme.txt file that is created in the gclas directory during installation on a Unix system.

A current version of GCLAS can be installed over an existing version. If that is done, the installer will present a warning indicating that the install directory already exists and requesting that the user verify that they wish to over write the existing files. Because more than one version of GCLAS can be installed on a computer at the same time, it is not necessary to overwrite older versions of GCLAS unless desired.

Getting Started with GCLAS

Starting the Program on a Computer Running the Windows Operating System

Installation of GCLAS on a computer running the Windows operating system by default will add "USGS" to the list of programs that you run from the Start menu. To start GCLAS,

- go to the **Start** menu and select Programs.

- drag down to **USGS** and select **GCLAS 1.05e** from the menu that pops up to the right.

A DOS window may open as GCLAS starts up. The GCLAS interface will appear shortly after that. You will probably want to maximize the GCLAS window when working with GCLAS.

Note: If the GCLAS window appears coarse or fuzzy, check the screen resolution (also called "screen area") to ensure that it is set to at least 1024 by 768 pixels. GCLAS requires a resolution of at least 1024 by 768 pixels for its display.

Starting the Program on a Computer Running the UNIX Operating System

With a standard installation of GCLAS on a UNIX machine, you should be able to start the program by typing the command **gclas**.

Importing Data for Analysis — A Short Example

Data can be imported into GCLAS from any local or networked file system. For purposes of getting started, however, a practice data set comes as part of the GCLAS installation. The following steps outline how to import one of the practice data sets:

1. From the File menu, select **Import** (or use the keyboard shortcut **shift-i**). This will open a file browser window.

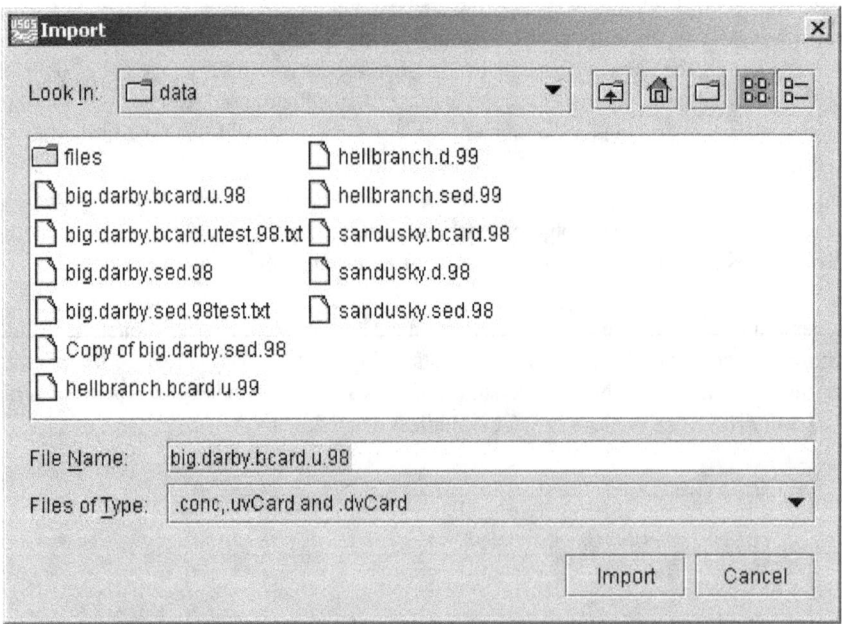

2. Using the **Look In** pick list at the top of file browser window, navigate to the data subdirectory of the GCLAS installation. In most PC-Windows installations, this will be on a local hard drive (probably C:\) in Program Files\USGS\gclas\data. With a standard UNIX installation, the practice data set can be found in the directory /usr/opt/wrdapp/gclas.

3. After you have located the data subdirectory, select the file big.darby.sed.98 (a card-image file containing sediment-concentration data) and left-click the **Import** button to load the concentration data into GCLAS.

4. Repeat the above process and select and load a streamflow unit-values file (big.darby.bcard.u.98; the concentration file and streamflow unit-values file must be from the same water year). Note: It is also possible to load daily values of streamflow in the daily-values card-image format.

5. Locate the folder and file-display tree in the left-hand panel at the top of the GCLAS window. Expand the tree by double left-clicking the folder icon that has a station number as its label or by single left-clicking the toggle icon to the left of the folder. You will see nodes (file names) for any sediment and streamflow data that have been loaded.

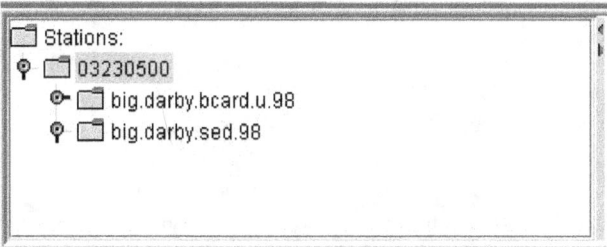

6. Now, left-click the folder icon associated with the concentration file. The name of the concentration node should be highlighted.

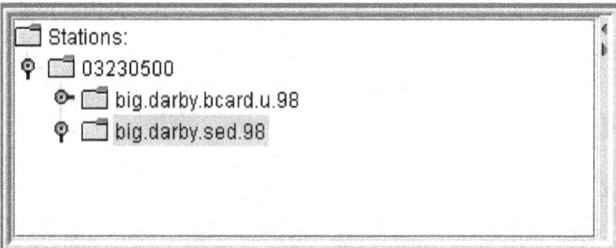

7. Click the right mouse button and choose the option to **Create/Edit GCLAS year**.

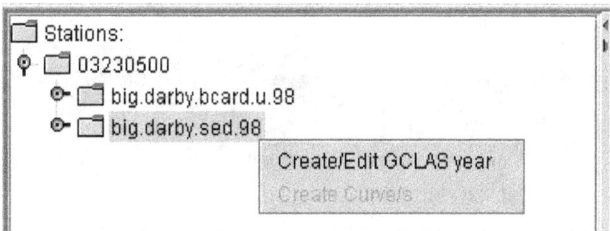

8. A window will appear (as shown below), and the dialog box at the bottom of this window will explain how to proceed. Either right-click in the plain white area of the window then select **Create/Edit GCLAS year** or left-click the button labeled **Create/Edit GCLAS year**.

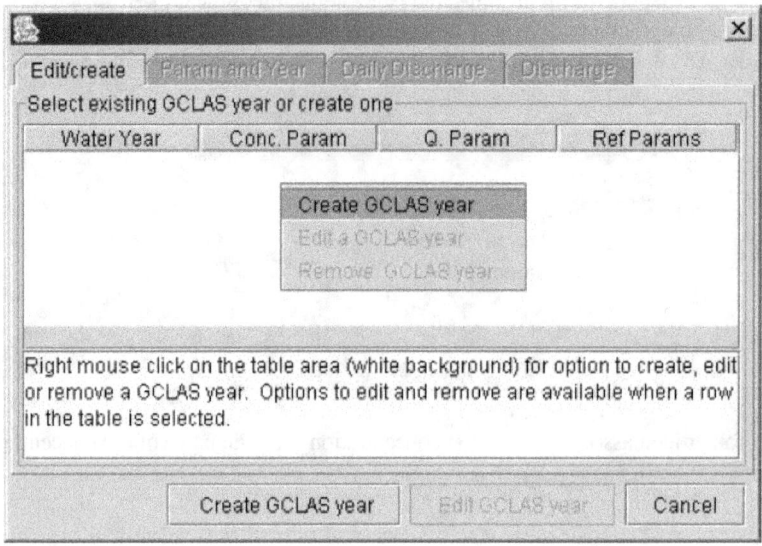

9. The **Param and Year** tab in the window will become active, as shown below. left-click the concentration parameter that you intend to work with (it will be highlighted when selected) and select the water year and load[3] unit. After the parameter, water year, and load unit have been selected, left-click the **Next** button.

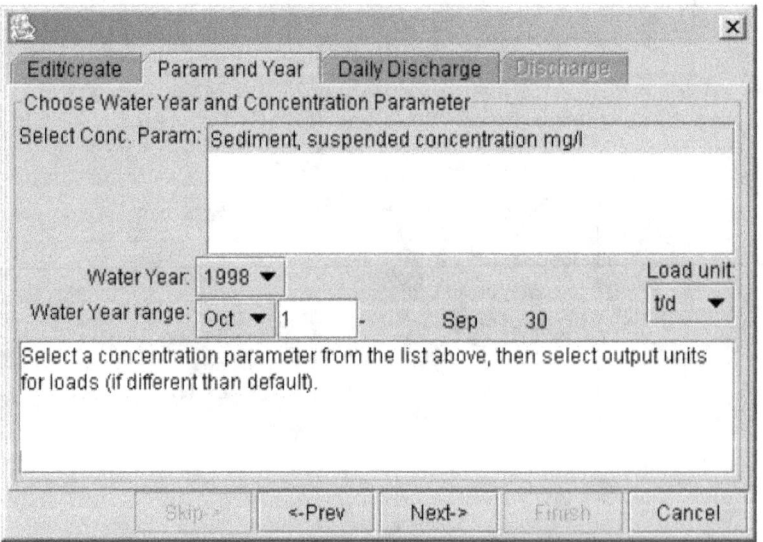

_____

[3]The term "load" refers to the mass or volume of material that is transported in a stream without implicit consideration for the rate of transport. The mass or volume rate of transport (for example tons per day) is actually a discharge. Because it is common to discuss the load for a given time period (for example the load for a day) and because it is also common to refer to streamflow as discharge, the term "load" will be use in this report and in GCLAS to refer to either the mass or mass rate of transport, and the term "discharge" will be used solely to refer to streamflow.

10. The **Daily Discharge** tab will become active as shown below. If no daily discharge data have been imported prior to this step, then no data-set names will appear in this tab. Loading of daily-values data is optional. Daily discharge data are not used for purposes other than display. If one or more daily discharge data sets have been imported, you can select one of the data sets and parameters and then left-click the **Next** button. Otherwise, left-click the **Skip** button.

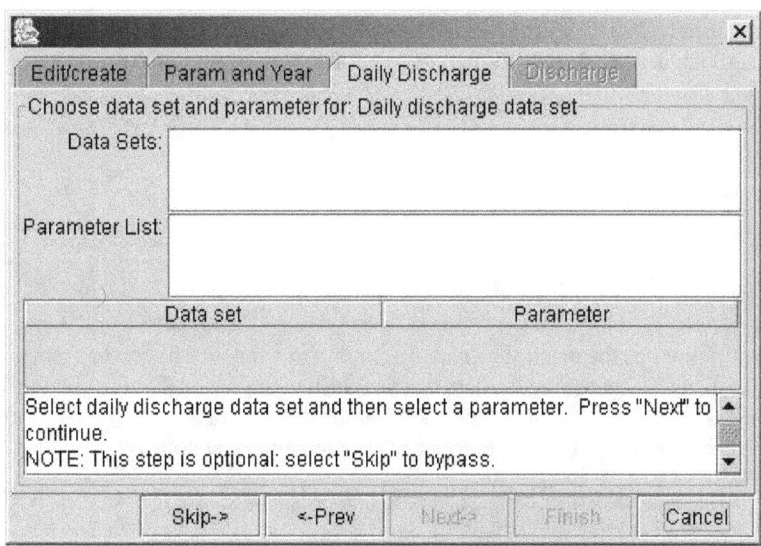

11. The **Discharge** tab will become active as shown below. Select the data-set name and parameter for the unit stream-flow data then left-click the **Finish** button. Once the **Finish** button has been selected, GCLAS will read the data sets and create the graphs and tables used during the GCLAS session. The time required to read the data sets and create the graphs and tables depends on the size of the data sets and on speed and memory capacity of your computer. Commonly, the time required ranges from about 5 to 30 seconds.

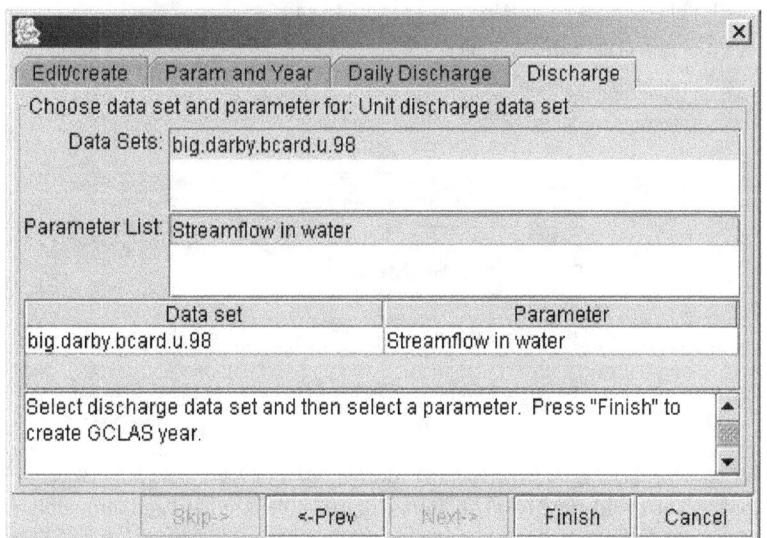

You are now ready to explore more features of GCLAS.

Importing/Opening Data Files

Import Command

The **Import** command (from the **File** menu; see below) is used to import plain-text (ASCII) data (either in card-image format or GCLAS tab-delimited "gcl" format) into GCLAS.

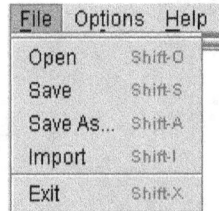

After you select **Import** on the menu shown above, an import window will pop up (shown below); from here, you can navigate to the directory containing the data that you wish to import. Make sure that the file type of the file that you wish to import has been selected from the pick list at the bottom of the import window (as in the example below) or else the file to be imported may not be displayed as a possible selection.

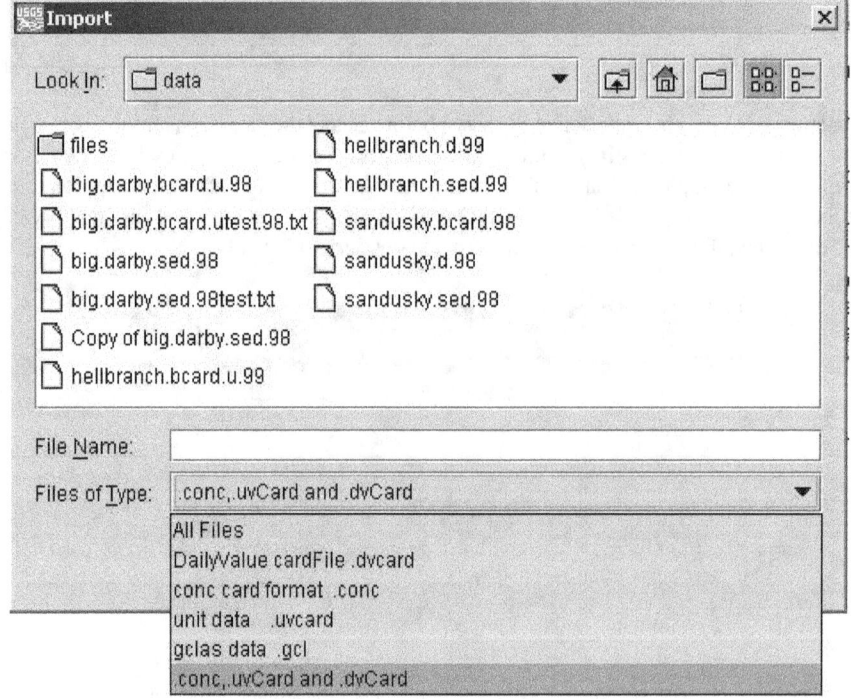

After a GCLAS year has been established (either by creating it or editing a GCLAS year that had previously been created), additional water-quality and (or) streamflow data can be imported and ultimately used as reference curves. References curves are discussed later in the section titled "Using Reference Curves." Data imported after a GCLAS year has been established do not have to be for the same water year as the GCLAS year.

Streamflow Data

GCLAS can read streamflow time-series data in the USGS National Water Information System (NWIS) daily-values and unit-values card-image formats. (See Appendix 4 for a description of these formats.) There is one exception to the unit-values card-image format described in Appendix 4. If the number of readings per day for streamflow is 1440 (corresponding to a 1-minute time interval), then a "Q" must be entered in column 31 of the first (and only the first) B-card image in the data set.

Water-Quality Data

GCLAS can read water-quality data in an ASCII-delimited "gcl" format and in the SEDCALC 2- and B-card image (SEDATA) format. (See Appendix 1 for description of SEDATA format.) The USGS Sediment Laboratory Environmental Data System (SLEDS) can output data in the 2- and B-card image format for entry into GCLAS.

At present, GCLAS is not designed to facilitate direct entry of measured concentration data from the keyboard, so all measured data must either be output from a USGS database or assembled in another application (such as SED-CALC, a spreadsheet, or an editor) and then imported into GCLAS. Microsoft Excel spreadsheet templates (called input_template.xls and input_template2.xls) for constructing concentration data files are available and are included as part of the GCLAS installation (in the folder or directory named "data"). The templates differ only in the input formats used for dates and times.

	A	B	C	D	E
1	# After completing form, press "Create gcl file" button to save data to Excel and gcl formats.				
2	# If macros are disabled, save file as tab- or comma-delimited file (as indicated by cell B10).				
3	#				
4	# creation date:	10/01/2003			
5	# creator:	jsmith			
6	#				
7	# header				
8	# station no:	03230500			
9	# name:	big.darby.sed.03	Create gcl file		
10	# separator:	tab			
11	#				
12	SampTime	SampTime_time	SampRepresentation	SampCollector	p80154Value
13	MM/dd/yyyy	HH:mm	id	id	val
14	10/01/2002	12:00	S	O	35.00

Figure 1. Sample Excel-based input template for creating GCLAS gcl input files.

Figure 1 shows part of a simple Excel-based input template, with line numbers on the left and letters designating columns on the top. Lines that begin with the "#" symbol in figure 1 are ignored by GCLAS with the exceptions of the station number, name, and separator, which are required entries. The station-number field is intended to contain an 8- or 15-digit station identifier. Station numbers are also present in other input-file formats, thereby facilitating the proper linkage of data from different files. The contents of the name field appears in the GCLAS file tree panel (discussed later) in association with the data contained in the file. Finally, the separator entry must be set to "tab" or "comma" to indicate the character that is used to delimit columns of data.

The keyword entries on line 12 of figure 1 identify the contents of data in the columns below. Figure 1 shows a small subset of the valid keyword entries. (Appendix 2 (p. 45) includes a complete list of keywords, formats/codes, and domains, where applicable.) Below each keyword entry is a format or code. The format or code dictates the format and (or) domain of the data that follow below in the column. The data must conform exactly to the indicated format and (or) domain for GCLAS to successfully read and interpret the data.

To use either spreadsheet template included with the GCLAS installation, do the following:

- **Open the spreadsheet template in Excel.** It is recommended that you make a copy of the original template in case you accidentally overwrite the template. Depending on your version of Excel and security settings, Excel may provide a security warning similar to that shown below:

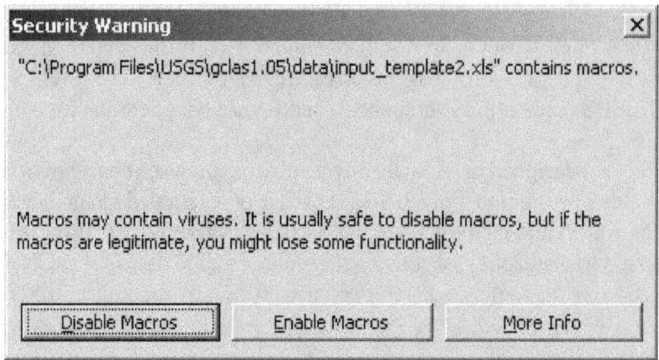

It is recommended that you enable macros. It is not essential that macros be enabled to use the templates, but enabling macros will make saving your input data easier.

- **Fill in the header data in column B.** Fill in creation date (mm/dd/yyyy format preferred), creator, station number, and file name in their respective rows, and choose between tab or comma delimited on the separator pick list. Note that the cells with red triangles in the corner will display instructions when you "mouse over" then (pause the cursor over them).

- **Fill in the body of the spreadsheet.** Fill in date and time (columns A and B in fig. 1) in the format specified on line 13. Enter the sample representation and sample collector using the pick-list options given. (Mouse-over instructions in the column headings list the options.) Enter numerical concentration values for each sample in the column labeled "p80154Value," where 80154 is the five-digit NWIS parameter code for suspended-sediment concentration. NWIS parameter codes for parameters other than suspended sediment also can be used. In addition to the standard NWIS parameter codes, generic parameter codes can be used within GCLAS (table 1).

Table 1. Generic parameter codes that can be used within GCLAS.

Parameter code	Description
99900	Generic concentration, mg/L (milligrams per liter)
99991	Generic concentration, mg/L (micrograms per liter)
99992	Generic discharge, t/d (short tons per day)
99993	Generic discharge, lb/d (pounds per day)
99994	Generic discharge, kg/d (kilograms per day)
99995	Generic discharge, Mg/d (megagrams per day)

If macros were enabled, once you have entered your data, you can save a copy of the Excel spreadsheet as well as a tab-delimited or comma-separated variable file (depending on the separator type specified in the header area of the Excel file) by left-clicking the **Create gcl file** button. The files that are created are named using the value contained in the name cell (cell B9 in fig. 1) as a prefix and ".xls" (for the Excel file) and ".gcl" (for the comma- or tab-delimited file) as the suffixes. If macros were not enabled for the input templates, once you have entered your data in Excel, left-click on **File** in menu bar and then left-click on the **Save as** option. A file window will open with a pick list titled "Save as type" at the bottom. Choose **Text (tab delimited)** or **CSV (comma delimited)** depending on the separator type that you entered in the header area of the Excel template file. Enter a name for the file (use the same file name that you specified in cell B9 of the Excel file), and then left-click on the **Save** button[4]. Unless you placed the file name in quotes (for example, "filename.gcl"), Excel will append an extension of ".txt" or ".csv" to the file, depending on the format in which the file was saved. The extension of the newly saved tab- or comma-delimited file (not the Excel file) should be renamed to ".gcl" if the file was saved with another extension. (Note: Some versions of Excel occasionally omit end-of-row delimiters on rows where there is no value in the last column. The omission does not occur in all cases, but GCLAS may not read the data properly if the delimiters are omitted. To avoid this problem, make sure that the last column of data is one that contains an entry for most rows.)

Open Command

The **Open** command (from the **File** menu) is used to open a previously created GCLAS project file. A GCLAS project file, which has an extension of ".gpf", is a special file format that GCLAS creates (using the **Save As** option on the GCLAS **File** menu); this file can contain all input data, coefficient information, and computed loads associated with a GCLAS session.

Once you have opened a GCLAS project file, you must select the data node corresponding to the concentration parameter from the file-tree panel (in the upper left corner of the GCLAS display). Select the node by left-clicking the toggles in the file-tree panel until the concentration node is visible, then left-click the concentration node, causing the node to be highlighted as shown in the example below:

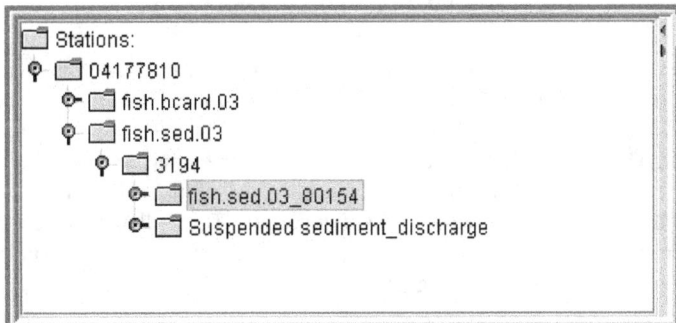

After the concentration node has been selected, click the right mouse button (with the cursor over the node) and choose the option to **Create/Edit GCLAS year** (example shown below) by left-clicking the option.

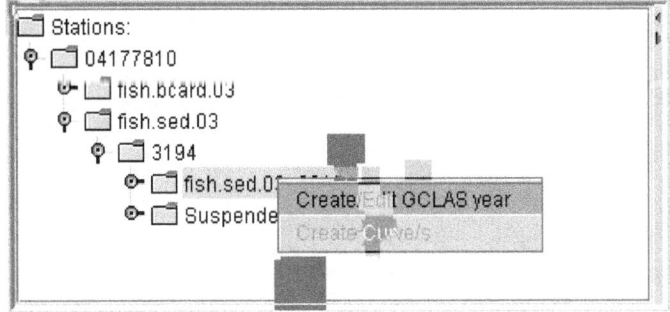

---

[4]It is also recommended that the file be saved in Excel format to facilitate future addition of or changes to data.

A new window will appear (example shown below) positioned to a tab labeled **Edit/create**. left-click the row corresponding to the water year that you wish to edit or view, then left-click the **Edit GCLAS year** button. Once that is done, the graph and tabular views of the selected data will be displayed, and the project data can be edited or viewed.

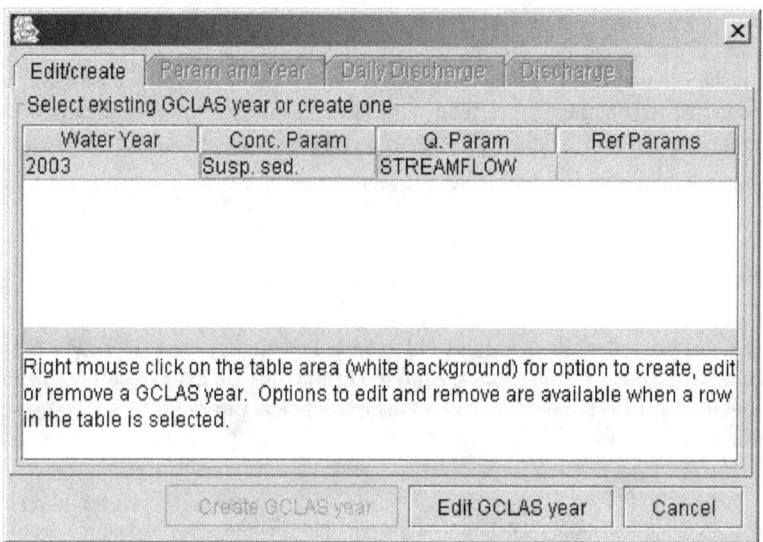

Overview of GCLAS Window and Panel Interface

Once data have been imported/opened in GCLAS and the GCLAS year has been created, you'll see a multipanel display showing the data in graphical and tabular formats. As a new user of GCLAS, you'll be confronted by two challenges: (1) learning the functions of all parts of the main GCLAS window and (2) learning to manipulate the various panels within the window, which differ slightly from other windows-type interfaces you may be familiar with. This section tackles both challenges in turn. Read this section carefully while looking back frequently at the initial layout of the main window. (**Do this before you resize or reposition panels.**) This will help prevent the attendant frustration of searching for missing panels hidden behind other panels. Also, maximize the GCLAS window to fit the whole monitor screen; this will minimize the need for resizing panels.

Organization of the GCLAS Display

Initially, seven panels are displayed in the main window. In addition, a window labeled "Ref. curves (Transport)" is minimized and shows as a small rectangle in the lower left corner of the main window.

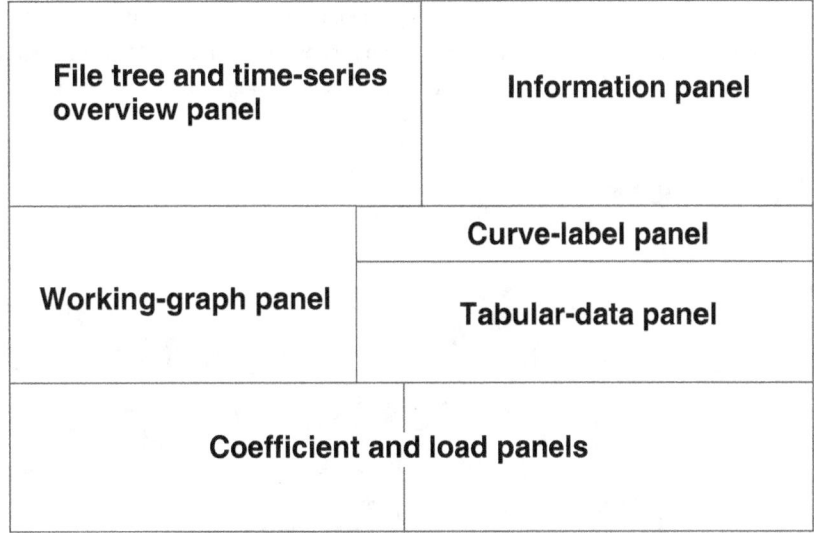

Figure 2. Schematic of GCLAS main window showing panel arrangement.

The **upper third** of the main window (fig. 2) contains two panels:

• At the left is the *file tree* and *time-series overview panel*. Three views of this panel are available by left-clicking on the icon tabs at the panel's right side. The *folder and file view* (top tab) shows the files that you've loaded into GCLAS and the folders that they're stored in, and it allows you to pick and choose the data to display and work on. The *graph view* (middle tab) shows an overview graph of your time-series data. The *information view* (bottom tab), when fully operational, will provide miscellaneous information about the GCLAS year being worked.

• At the right is the *information panel*, which presents startup information at the beginning and then shows information about specific data points when you're actually working with your data.

The **middle third** of the main window contains three panels:

• At the left is the *working-graph panel*. This panel is where estimated concentration data can be edited directly from a graph display by using the mouse to add and remove data points, shift curves, and so on.

• Two vertically stacked panels are at the right of the working-graph panel. The top one is the *curve-label panel*, which displays a color-coded list showing the relation between points and curves in the graph and the time-series data represented, plus information about attributes of the data being displayed and which elements of the display can or cannot be manipulated.

• Immediately below the curve-label panel is the *tabular-data panel*, which displays the combined streamflow and concentration data in tabbed tables. Changes made to the tabular-data display are immediately reflected in the working graph and vice versa.

An alternative configuration of the preceding three panels is available by left-clicking the panel orientation button in the upper right corner of the working-graph panel. Clicking this button changes the orientation of the working-graph and tabular-data panels from side-by-side to vertically stacked. The many tools and functions available in these panels are discussed in the section titled "Adding and Editing Water-Quality Data" on page 19.

The **bottom third** of the main window, which is used for computing and applying cross-section coefficients and computing loads, contains two panels whose functions change, depending on which of the tabbed folder views (**Calculate Coefficients**, **Apply Coefficients**, or **Compute Loads**) has been selected. Details about this panel are given in the sections titled "Analyzing and Applying Cross-Section Coefficients" and "Computing Loads."

Transport-Relation Window

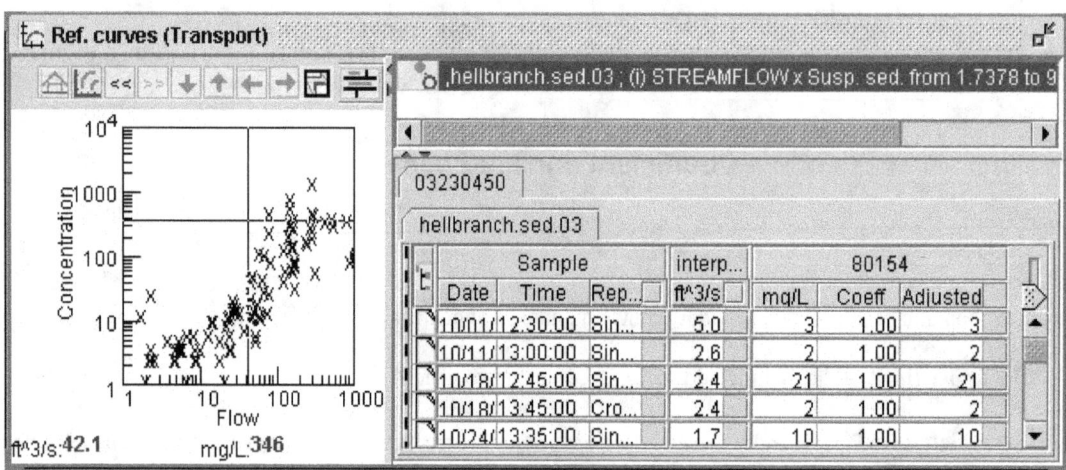

Figure 3. Transport-relation window.

When maximized, the *transport-relation window* (fig. 3) shows a scatterplot of concentration as a function of streamflow for all non-estimated concentration data in the input data set that can be used to help estimate and fill in periods with missing data. Double left-click anywhere on the minimized form of this window (a small rectangle initially in the lower left corner of the GCLAS main window) to maximize it. To minimize it again, left-click the small square with an arrow pointing to it (upper right corner of the maximized window). Details about use of this window are given in the section titled "Using the Transport-Relation Window to Aid Estimation" on page 27.

Working with GCLAS Panels, Tabs, and Buttons

The GCLAS windows (main window and transport-relation window) work similarly to windows on a PC or UNIX machine. You can, for example, drag them around on your computer screen, and you can minimize them by left-clicking on a small box in the upper right corner of the frame. Panels, however, behave differently. You cannot drag them around the screen like you can windows, but you can resize them in two ways:

- Left-click any edge of the frame of the panel and drag that edge to expand or contract the panel by as much as you wish, up to the panel's horizontal or vertical limit. You will know when the cursor is over a draggable edge because it changes to a horizontal, vertical, or slanted double-headed arrow. The orientation of the double-headed arrow indicates the directions in which the edge can be dragged.

- Left-click the small arrows on the frame ╎ to maximize the panel within its section of the main window (outward-facing arrow) or return the panel to its former size (inward-facing arrow).

If a panel disappears when being resized, it most likely is hidden behind another panel. If that happens, uncover the hidden panel by sequentially resizing panels that normally are adjacent to it. Once you get the hang of the panel arrangement, you'll find that it's convenient to hide unused panels by sliding panels of current interest in front, thereby maximizing your active working space.

Tabs are another important feature of certain GCLAS panels. They look similar to index tabs in a three-ring notebook or on manila folders. You can use tabs to locate and display different views of a multipart panel.

Buttons, which are used to execute common functions, are included at the top or side of some of the panels, and they work much like buttons do in other windowed software.

Adding and Editing Water-Quality Data

Analyzing and editing water-quality data primarily involves the panels in the upper two-thirds of the main GCLAS window (fig. 2). Data shown in most graphs and tables are dynamically linked so that modifications to data in one panel are automatically reflected in the other panels. For example, if an estimated concentration value is modified in the tabular-data panel, its position in the working graph is automatically updated, and vice versa. The curve-label panel also is synchronized with the working-graph and tabular-data panels, helping you distinguish between curves and showing what data manipulations are enabled or disabled.

Using the Overview Graph and Working Graph to Display Curves and Data Points

On the top left panel is a tab with a graph icon ⌨. left-click on this tab and you will see an overview graph showing a streamflow curve (in blue) and a concentration curve (in red) for the GCLAS year. When you see the overview graph, move the cursor into the white graph area and press and hold the left mouse button; then drag the mouse to create a box that spans at least 15 days on the graph (fig. 4). When you release the mouse button, the working graph will be zoomed in, and a box showing the extent of the zoomed view will be displayed (fig 4). Repeat this procedure if you wish to see a different part of the GCLAS year in the working graph

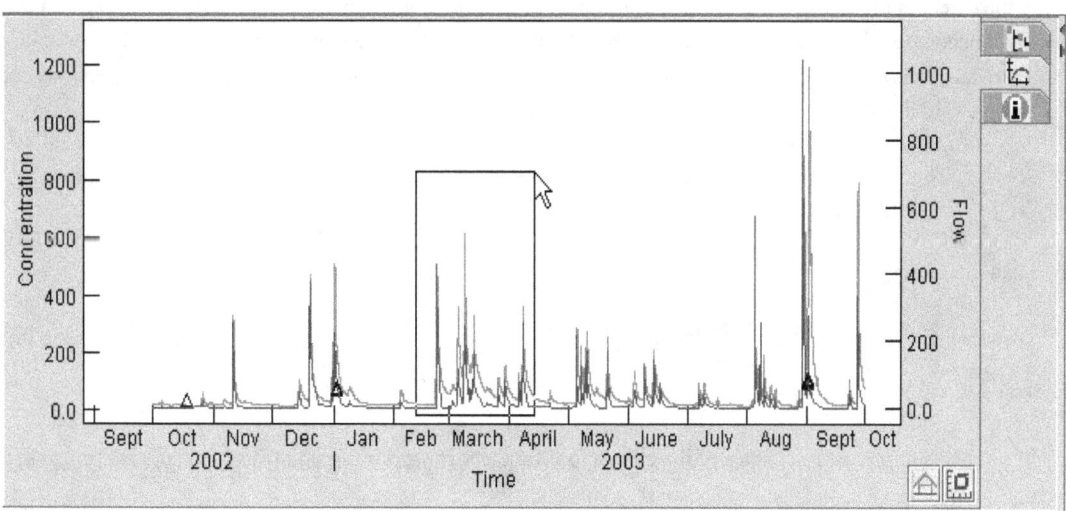

Figure 4. Time-series overview panel.

Note the two buttons at the bottom right of the overview panel (fig. 4). The one on the right, with the box-and-axis icon 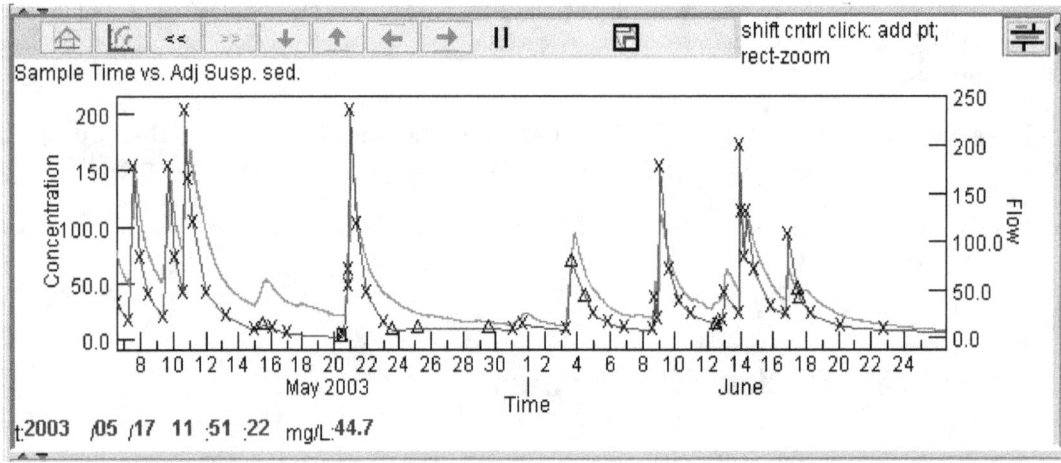, lets you zoom the overview graph to the currently defined zoom box. The one on the left, with the home icon 🏠, takes you back to the default water-year view.

Figure 5. Working-graph panel.

Note the shape of the symbols in the working graph:

- A *rectangle* means that the data point is for a sample whose concentration is representative of the mean concentration in the cross section (for example, a sample collected by integrating over the width and depth of the stream).

- An *X* means that the data point is an estimated value.

- A *triangle* means that the data point is from a sample collected at a point or single vertical.

The zoom features operate within the working graph in a fashion similar to that of the overview panel. If you move your cursor into the working graph, then press and hold the left mouse button and drag to create a box, the working graph will be zoomed to include the area bounded by the box, and the overview graph will update to display a box showing the extent of this zoom. Additionally, a button on the toolbar immediately above the working graph with a << symbol will become enabled. left-click on the << (back) button to return to the previous zoom view. Now, left-click on the >> (forward) button to return to the newer zoom view. GCLAS allows you to go back to multiple previous zoom views and forward again.

If you wish to pan the GCLAS curve from the current zoom view, use the arrow buttons on the toolbar ⬇ ⬆ ⬅ ➡ to shift the working-graph view vertically or horizontally. The arrow keys on the keyboard can also be used to pan the current zoom view.

If you wish to go back to the default water-year view, use the button at the extreme left of the toolbar, with the home icon 🏠.

If you wish to rescale the y-axes of the curves to fill the entire working-graph frame, use the maximize scale toolbar button 🗠. The individual y-axes (corresponding to different parameter or units) cannot be scaled independently of one another.

Adding/Editing Estimated Values in the Working-Graph Panel

To add estimated values in the working-graph panel, position the cursor to the desired location within the graph and then, while holding down the shift and control keys, click the left mouse button. The x and y coordinates of the cursor are tracked continuously in the working graph and displayed in the lower left corner of the panel. (See fig. 5.) The coordinate information is particularly useful when you're graphically adding estimated values to help ensure that estimates are added at or near the desired concentration and time coordinates. If after adding an estimated data point you determine that it is not at the intended concentration or time coordinate, you can modify the point in two ways:

- *Remove it.* Click on the point with the left mouse button to select it, then click on the point with the right mouse button to bring up a small two-item menu and select **Remove Pt(s)**.

- *Modify its location.* Left-click on the point and drag it vertically (up or down) in the working graph or edit the time or concentration in the tabular-data panel. Time coordinates of a point cannot be edited by dragging; however, they can be edited in the tabular-data panel. To edit time or concentration values of estimated data in the tabular-data panel, you must double left-click in the cell to begin editing and press the **Enter** key to register the edit.

Extra care should be used when adding estimated values at concentrations near zero because negative concentrations can result if the mouse is slightly out of position when its button is clicked to register the value. GCLAS does not prevent entry of negative values, nor does it give any warning of their presence.

Left-click a data point in the working-graph panel to find and highlight the corresponding data-point entry in the tabular-data panel and vice versa. This feature can be useful when two points of interest plot very close to each other, too close to effectively use the mouse to select both; you can select one of the points on the graph, then use the table to select the other point.

The text that appears to the right of the button bar (fig. 5) contains hints for use. The phrase "shift cntrl click: add pt" is intended to remind the user that an estimated value can be added by holding down the shift and control keys on the keyboard while left-clicking in the working graph. Similarly, the phrase "rect-zoom" is intended to be a reminder that you can zoom in the graph windows by holding down the left mouse button and dragging to enclose the intended zoom extent within the rectangle that is drawn as the mouse is moved.

Setting the Default Representation Type of Estimated Values

When estimated values are added, they are assigned a representation type of "cross section," "single vertical," or "point." The representation type is used to determine which, if any, cross-section coefficients can be applied to the value. Samples with a representation type of "cross section" are assumed to be representative of the mean concentration of that constituent within the stream cross section. As a consequence, cross-section coefficients (other than 1.0) are never applied to samples with a representation type of "cross section." Samples with a representation type of "single vertical" or "point" are not assumed to be representative of the mean concentration of that constituent within the stream cross section. Sample concentrations with these representation types can be modified by application of coefficients of the identical type. A single data set can contain samples of one, two, or three representation types.

By default, GCLAS assigns a representation type of "cross section" to newly estimated values; however, the default representation can be changed for the duration of a GCLAS session by choosing a different representation from the **Options** menu shown below:

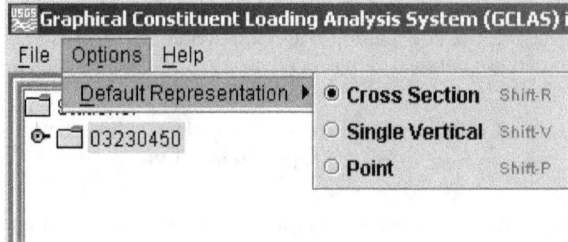

Left-click the radio button next to a representation type to change the default representation type. The change in default representation type will affect only those estimated values that are added subsequently within the current GCLAS session; the representation type always defaults to **Cross Section** at the start of a GCLAS session. Any estimated values that were added before the change in default representation type will retain the representation type assigned at the time they were added. If you wish to change the representation type of an estimated value that already has been added, you must follow the instructions given below for editing data in the tabular-data panel.

Editing Data in the Tabular-Data Panel

The tabular-data panel (fig. 6) displays data in a tabbed-folder format with different types of data and data from different sites organized in separate tabbed tables.

	Sample			interp:00060	80154		
	Date	Time	Representa...	ft^3/s	mg/L	Coeff	Adjusted
	04/27/2003	20:05:16	Cross Secti...	8.9	E165	1.00	165
	05/01/2003	13:10:00	Single Verti...	10	3	1.00	3
	05/05/2003	6:18:45	Single Verti...	33	E6	1.00	6
	05/05/2003	10:20:55	Single Verti...	100	E150	1.00	150
	05/05/2003	11:45:00	Single Verti...	155	255	1.00	255

03230450 — hellbranch.sed.03_80154 — hellbranch.bcard.03

Figure 6. Tabular-data panel.

One of these tables has a tab with the concentration file name followed by the parameter code for the constituent of interest (for example, hellbranch.sed.03_80154 as shown in fig. 6). This table—referred to as the "GCLAS water-quality table" hereafter—contains both concentration data and streamflow data (machine interpolated, if necessary), and it is editable. The GCLAS water-quality table is divided into three or more sections that are organized under spanner headings. For example, the spanner headings shown in figure 6 are labeled **Sample**, **interp:00060** (for interpolated streamflow), and **80154** (the parameter code for suspended-sediment concentration). Under each spanner heading is one or more columns containing data or metadata. Metadata may be important for some analyses, but it can be a distraction for others. Consequently, GCLAS was designed so that you can toggle between detailed and less detailed views of the data by left-clicking on spanner and (or) column headings. For example, after left-clicking the **80154** spanner heading shown in figure 6, the spanner heading expands to show all of the columns contained within it (fig.7), and the name on the spanner heading changes to reflect the text description corresponding to parameter code 80154. For additional data-viewing flexibility, you can reorder columns by dragging a column heading to another position within the same spanner heading. You can also use the scroll bars to navigate through the table and the slider tool [image] (near the upper right corner of the GCLAS water-quality table) to adjust the spacing between rows for easier viewing.

Sediment, suspended concentration mg/l								
mg/L	Remark	Usable	Status	Type	QA Type	Coeff	Adjust...	
E165		✔	Presumed...	Lab	Regular	1.00	165	
3		✔	Presumed...	Lab	Regular	1.00	3	
E6		✔	Presumed...	Lab	Regular	1.00	6	
E150		✔	Presumed...	Lab	Regular	1.00	150	
255		✔	Presumed...	Lab	Regular	1.00	255	

Figure 7. Tabular-data panel showing expanded spanner heading.

When a constituent such as sediment is selected for working a record, subcolumns for cross-section coefficients and the adjusted concentrations that result from the application of the coefficients (the last two columns shown in fig. 7) are added to the table, and the curve that is plotted will reflect the coefficient-adjusted concentrations.

The other tabbed table contains streamflow data as contained in the original card-image file, and its tab label reflects the name of that data file; this table is for reference only and is not editable. To switch between tables, left-click on the appropriate tab.

The GCLAS water-quality table contains a wealth of ancillary information (metadata) associated with each data value. Some of these metadata may have been placed in the table as default entries when external files were read in and the GCLAS year was created. Most of these entries, however, can be changed to match your knowledge of actual conditions or can be set to accurately represent added data. Some of the columns in the GCLAS water-quality table are self-explanatory, but others, especially those that have pick-list options, require some description.

Under the **Sample** spanner heading:

- **Representation** (4 options). Sample-representation type includes cross section, point, single vertical, and unspecified.

- **Hydro Condition** (7 options). Hydrologic condition includes choices for stable flow at various flow levels or for rising or falling stage.

- **Collector** (4 options). Hydrologist, observer, automatic, unspecified.

- **Hydro event** (12 options). Hydrologic event choices, including hurricane, flood, storm, routine sample, and others.

- **Method** (7 options). Sample-collection method includes equal discharge increment (EDI), equal width increment (EWI), pumped, single vertical, grab, unspecified, and none.

- **QA Type** (12 options). Quality-assurance type includes replicate, duplicate, composite, spike, and others.

Under the **80154** (or comparably named concentration parameter) spanner heading:

- **Remark**. The remark field is not currently implemented.

- **Usable** (check box). This column allows you to stipulate that a value should not be used (by unchecking it). Unused values will continue to be displayed in the graphs and tables but will be ignored for all other purposes.

- **Status** (5 options). The status field shows information about the quality-assurance status of the datum, consisting of the following options: awaiting review, presumed OK, accepted, rejected, and not reported

- **Type** (2 options). The type field shows information about the source of the measurement with options of field or lab.

- **QA Type** (12 options). Same options as under **Sample** spanner heading.

Data in the GCLAS water-quality table can be manipulated in three ways:

- *Numerical data* can be changed by double left-clicking on the table cell to enable editing. Position the cursor with a single left click, then delete, backspace over, or type in new numbers. Double left-click to highlight all data in the cell to delete or type over everything. Then press the **Enter** key to apply the change.

- *Pick-list data* can be changed by double left-clicking on the pick-list table cell, left-clicking on the desired entry from the pick list, and then left-clicking in any other table cell on the same line in the table and under the same spanner heading as the cell being edited. Alternatively, a pick-list selection can be registered by pressing the **Enter** key while holding down the **Shift** key.

- *Date and Time data* can be changed in a fashion similar to other numerical data; however, certain special editing characteristics are associated with these fields. Date and time editing functions are described in detail in Appendix 3 (p. 48).

- *Check boxes* can be checked or unchecked by left-clicking twice and then pressing **Enter**.

As you change concentration or discharge values in the table, the curves and symbols in the working graph and the overview graph will reflect the changes accordingly.

Using the Curve-Label Panel

The curve-label panel (fig. 8) shows information about what data are being displayed and which elements of the display can or cannot be manipulated. It also is used to change settings that enable or disable certain editing features. Features of the curve-label panel are described below.

Figure 8. Curve-label panel.

Curve text label:

- *Color coding* of curve labels corresponds to that of the curves themselves.

- *Line through label name* indicates that the corresponding data set is not displayed on the graph.

- *Order of curve labels in the list* indicates and controls the order in which curves are plotted; the curve listed first is plotted on top. Reordering curve labels changes the plotting order. This can be done by left-clicking on a label and dragging it. (The cursor will change to a different symbol while the curve label is being dragged.)

Symbols on icons to the left of the label:

- Ø *Circle and slash* means that the curve will not respond to manipulation with the mouse. Absence of the circle and slash means that mouse manipulation is enabled.

- ◆ *Red ("active") dot* means that the curve is the active curve. The coordinates that are tracked in the working graph are those of the active curve.

- **I** *I-beam* symbol means that the curve is editable. Absence of I-beam means that the curve cannot be edited.

- ○ *Point symbol* (an open circle in the lower right corner) means that symbols are displayed on the working graph.

- ╱ *Line symbol* (a forward slash in the upper right corner) means that lines (curves) are displayed on the working graph.

- *Carets* indicate whether a curve can be moved ⌃ or has been moved ⌃̂. A slash through the caret ⌃̸ indicates that the curve has been moved but cannot presently be moved.

Pop-up menu accessed by right-clicking curve label text (fig. 9):

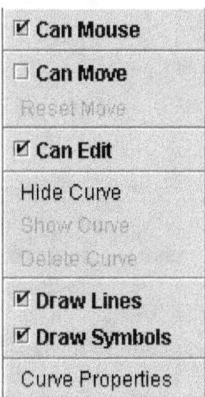

Figure 9. Curve label pop-up menu.

- **Can Mouse** enables or disables "mousing" (data-point manipulation with the mouse) for the curve of interest. For example, you might wish to disable mousing of a streamflow curve so that you can better manipulate a concentration curve if the two curves are close together.

- **Can Move** enables or disables curve movement with the mouse. When enabled, the curve can be moved by pressing the control key and simultaneously left-clicking on a point on the curve and dragging with the mouse. (The middle mouse button also works on some UNIX machines.)

- **Reset Move** restores the curve to its original position (and thus serves as an undo for a move).

- **Can Edit** allows a curve/data set to be edited.

- **Hide Curve** removes a previously displayed curve from the working and overview graphs.

- **Show Curve** displays a previously undisplayed curve on the working and overview graphs.

- **Delete Curve** eliminates the entry in the curve label list (so you'll need to recreate the curve from the file tree if you wish to bring it back).

- **Draw Lines** plots a line between the points, colored to correspond to the color of the curve label.

- **Draw Symbols** plots the measured data points as triangles or rectangles and estimated data points as X's.

- **Curve Properties** — this feature is not yet available.

Grayed-out entries in the pop-up menu cannot be selected.

Using Reference Curves

Water-quality and (or) streamflow data that have been imported during a GCLAS session can be used as reference curves. Reference curves are time series that are used for comparative purposes. For example, turbidity time series are sometimes used to help estimate the shape of a sediment-concentration time series during periods when concentration data are missing or undersampled. Once a GCLAS year has been established, other time series that have been imported during the GCLAS session can be used as a reference curve by left-clicking on the corresponding node for the time series in the file-tree panel and then left-clicking on the **Create Curve/s** button as shown below.

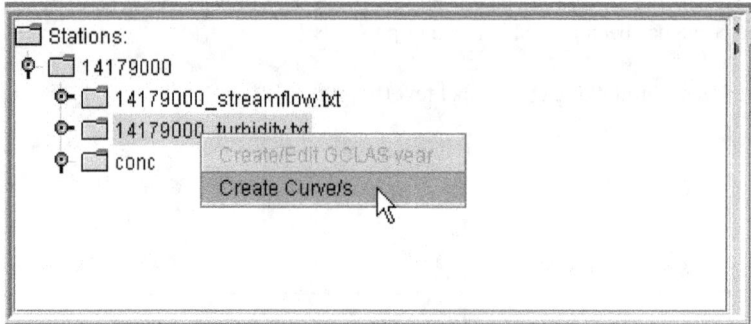

The parameter and water year to be used for the reference curve must be selected in the window that pops up. As shown in the example below, left-click on the **Create Curves** button to finish creating the reference curve. The water year of

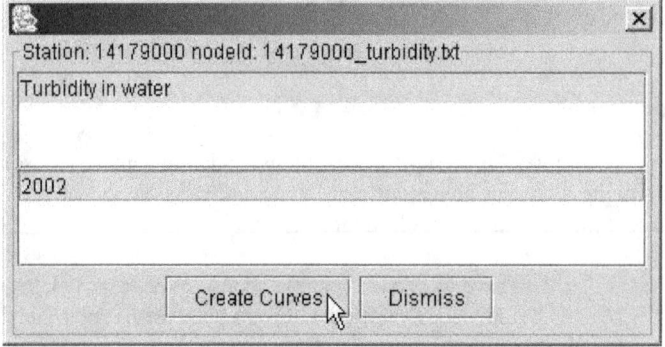

the reference curve does not have to be the same as the GCLAS year. Irrespective of the water year chosen, the reference curve will always be plotted on the same timescale (ignoring the difference in years) as the GCLAS year. After the **Create Curves** button has been clicked, the reference curve will be displayed in the overview and working-graph panels. In addition, the reference curve will become the active curve. In general, you will want your primary concentration time series to be the active curve. In that case, double left-click on the entry in the curve-label panel for the primary concentration time series to make it the active curve. The active curve can be identified by a red dot in the upper right-hand corner of the icon that appears to the left of the curve label in the curve-label panel.

Using the Transport-Relation Window to Aid Estimation

To compute accurate loads, the constituent time series must be drawn accurately. Most commonly, you'll need to estimate data for periods when measured data are lacking. Developing those estimates requires knowledge of the stream hydrology and of the transport characteristics of the constituent of interest.

The transport-relation window (fig. 3) shows a transport plot—a plot of concentration as a function of streamflow for data in the data set. GCLAS displays crosshairs in the transport-relation window that intersect at coordinates that are a function of the position of the cursor in the working-graph panel. Specifically, the streamflow (x) coordinate is set to the streamflow that is coincident in time with the position of the cursor in the working-graph panel, and the concentration (y) coordinate is set to the concentration value corresponding to the position of the cursor in the working-graph panel. The crosshair display facilitates estimation of missing values by allowing you to consider ancillary factors and cues (such as observed recession characteristics) and at the same time see how well alternative estimates fit with previously observed transport characteristics.

GCLAS displays a single transport curve showing only unadjusted measured (not estimated) concentrations in the input data set. The fact that concentration data in the transport-relation window are not adjusted (that is, they do no reflect application of any coefficient relations) has important implications for its use. If the majority of concentration values in the data set are derived from samples collected at a single point or vertical, then it is best to use the transport-relation window for estimation before applying coefficient relations. If instead, the majority of concentration values in the data set are derived from samples collect by means of depth- and width-integrating techniques (such as EWI or EDI; see Edwards and Glysson, 1999), then coefficient relations should be applied before using the transport-relation window for estimation.

If the input concentration and streamflow data files contain more than one water year of data, then the transport curve will show measured concentrations for all years over the multiyear range of flows. (However, with the exception of reference curves and reference-curve data, the overview graph, working graph, and tabular-data panel will reflect only the water year being actively worked.)

The panel that contains the transport curve has basically the same layout as the working-graph panel. The buttons have the identical functions, and the display of coordinates shows the value of flow and concentration at the intersection of the crosshairs if the cursor is in the working graph or of the cursor if the cursor is in the transport-curve graph.

You'll see two other dynamically linked panels in this window:

- The *transport-curve curve-label panel* looks like the working-graph curve-label panel, but the labels describe the data shown on the transport curve and the range of values to which they apply. The icon symbols are the same as those in the curve-label panel, and an identical pop-up menu appears by right-clicking on the curve name.

- The *tabular transport-data panel* lists the numerical data that are represented graphically in the transport curve. These data can be exported for statistical analysis outside GCLAS. (A right click on the table brings up the **Report** option, which allows you to write data displayed in the table to an rdb file.)

Analyzing and Applying Cross-Section Coefficients

In essence, cross-section coefficients are used to correct for systematic bias in concentrations measured at a single point or vertical as compared to the true mean concentrations in the cross section. Cross-section coefficients commonly are determined from concentrations for one or more samples collected at a single point or vertical and one or more samples that are integrated over the width and depth of the cross section. Coefficients should be determined only from sets of samples that are collected close in time to each other and at streamflows that not appreciably different from one another. In the simplest case, the coefficient is computed by dividing the concentration obtained from a depth- and width-integrated sample by the concentration obtained from a sample collected at a point or single vertical. After cross-section coefficients have been determined from sample sets collected from a representative sampling of seasons and streamflows, the analyst can view the coefficients as a whole to look for trends with respect to time and (or) streamflow. More detail about the computation of cross-section coefficients and their subsequent analysis can be found in Porterfield (1972).

Construction of cross-section coefficient relations can be complex. GCLAS includes a variety of tools to aid in (a) computing the cross-section coefficients, (b) assessing trends in the coefficients, and (c) visualizing and applying coefficient relations. In addition, GCLAS can track, store, and apply cross-section coefficients that apply to both single-vertical and point samples concurrently. Consequently, separate cross-section coefficient relations can be determined and applied, for example, in those instances where samples are collected by an automatic sampler and at a single vertical by an observer or technician.

Selecting Samples and Calculating Sample-Based Coefficients

1. Using the working graph in combination with the tabular-data panel, locate a depth- and width-integrated (cross-section) sample and any point sample(s) or single-vertical sample(s) that you wish to relate to it. As with any calculation of cross-section coefficients, the samples to be related should have been collected relatively close together in time and at approximately the same flow.

2. Select one of these samples by left-clicking either the symbol in the working graph or the corresponding row in the data table.

3. Select the other sample(s) by holding down the control key while left-clicking (or holding the shift key and left-clicking to select a range of samples).

4. Now look at the lower left panel in the main window. If the **Calculate Coefficients** tab is not already selected (as indicated by a darkening of the tab relative to the others), then left-click the tab to select it.

5. Left-click the **Get** button to bring the selected samples into this panel.

6. Left-click the **Calc.** button to calculate the coefficient. The white boxes in the lower part of the panel will show the results and indicate the coefficient type (point or single vertical) as shown in the example below (fig. 10).

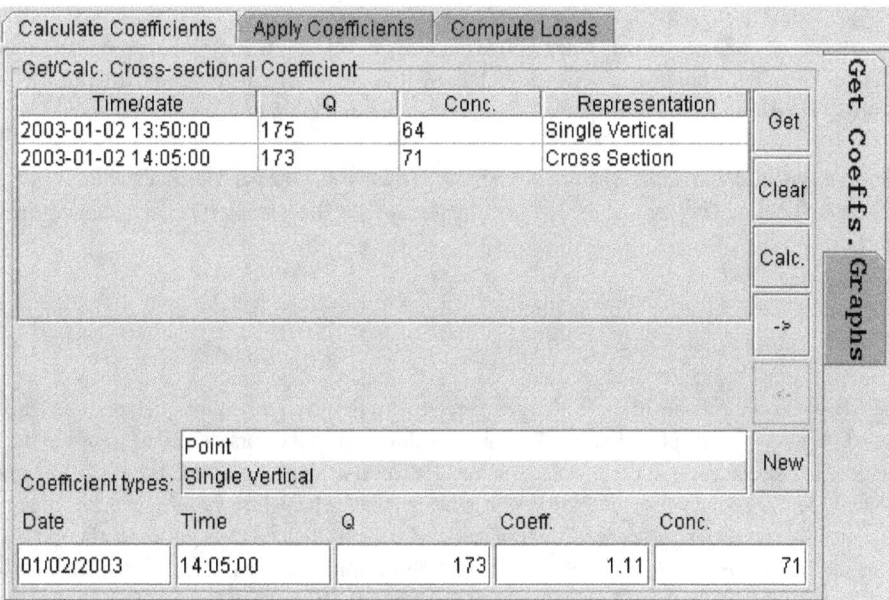

Figure 10. Get Coeffs. tab of the Calculate Coefficients panel.

7. Left-click the right-arrow button (**->**) to move the computed coefficient and supporting data into the holding area in the upper part of the adjacent panel. The illustration below (fig. 11) shows the entry above after it has been moved to the holding area (the upper part of the panel).

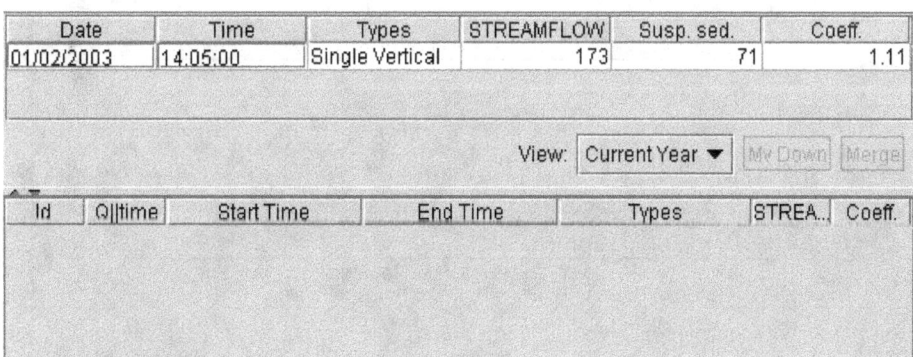

Figure 11. Coefficient holding area (top) and coefficient relation area (bottom) of the Calculate Coefficients panel.

8. Repeat the above procedure for as many more sample sets as you wish to work with for the current GCLAS year.

Notes:

*   If you can see only one row of data when you're placing more than one coefficient computation in the holding area, either enlarge the holding-area panel slightly by dragging down the bottom border or use the scroll bar (which will appear whenever undisplayed rows of data are in the holding area) or do both.

*   If you wish to inspect the original sample values used to construct the coefficient data, left-click on the row of interest in the holding area and use the left-arrow button (<-) in the lower left panel to display the original sample data there.

*   To delete a coefficient placed in the holding area, right-click the row of interest in the holding area and then left-click the **remove** option in the pop-up menu.

*   To view how the coefficients in the holding area related to flow and time, left-click on the **Graphs** tab in the left-hand panel to see a display (fig. 12). This is a good way to spot errors as well as trends in the coefficients. If you need to correct an error, (a) left-click on the **Get Coeffs.** tab, (b) left-click the erroneous row in the holding area, (c) left-click the left-arrow button (**<-**) to bring the data back into the editing area, (d) correct the data, recalculate, and then send it back to the holding area with a left-click on the right-arrow button (**->**). The curve-label panels on the **Graphs** tab (fig. 12) work in a fashion similar to that of the curve-label panels discussed earlier. In particular, it may be useful to turn off drawing of the lines (by right-clicking on the *Coeff.* entry and unchecking the **Draw Lines** box on the pop-up menu) to help visualize trends in coefficients.

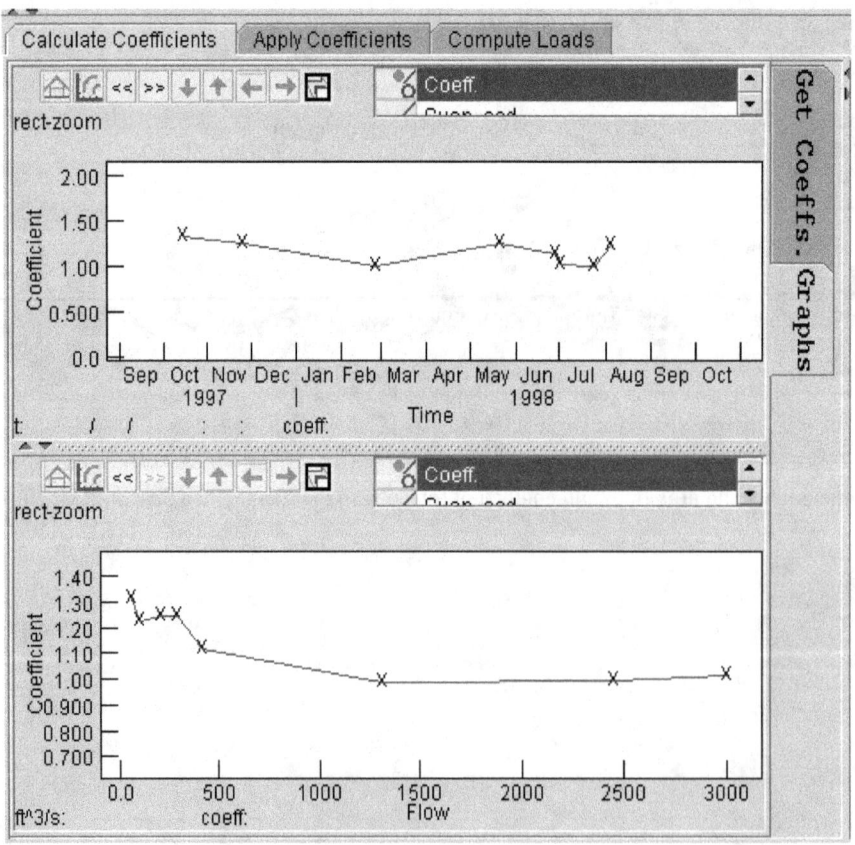

Figure 12. Graphs tab of the Calculate Coefficients panel.

9. After you've calculated all your coefficients for the GCLAS year and have moved them to the holding area, you have two choices before beginning to apply them to the record:

   - Move coefficients individually into the coefficient-relations area below the holding area by means of the **Mv Down** button. Coefficients that are moved into the coefficient-relations area also will appear in the *coefficient-relations panel* (located on the **Apply Coefficients** tab), where they can be used to create coefficient relations.

   - Combine (average) similar coefficients by means of the **Merge** button; the merged data will be moved as a single row into the coefficient-relations area below the holding area. Note that the start time will be the earliest date and time among the coefficients merged, and the end time will be the latest; the streamflow will be the rounded average flow of the entries being merged, and the coefficient will be their average.

Direct Entry of Coefficients

If you already know what coefficients you want to apply and don't need to calculate them in GCLAS, do the following:

   - Left-click the **Calculate coefficients** tab if it hasn't already been selected.

   - Left-click the **New** button.

   - Select either Point or Single Vertical in the **Coefficient types** box.

   - Enter coefficient data in the five boxes at the bottom of the panel. (See fig. 10.)

   - Press the right-arrow button (**->**) to move the new coefficient to the holding area at the right; then see step 9 in the section titled "Selecting Samples and Calculating Sample-Based Coefficients."

Applying Coefficients as a Function of Streamflow

Use the following steps if you wish to adjust concentration data using a coefficient relation that varies as a function of streamflow:

1. Left-click the **Apply Coefficients** tab in the lower left panel after you've moved one or more coefficients down into the coefficient-relations area. The coefficients that were moved in the coefficient-relations area will be listed in the coefficient-relations panel.

2. Left-click either the **Point** or the **Single Vertical** tab in the *Application of coefficient relations panel* (depending on which type of coefficient you wish to work with), then left-click on the **New QC** button (to create a new relation between streamflow and concentration). Immediately after clicking the **New QC** button, the panel labeled *Q's influ-*

ence on Coeff.s will change to look something like that shown below.

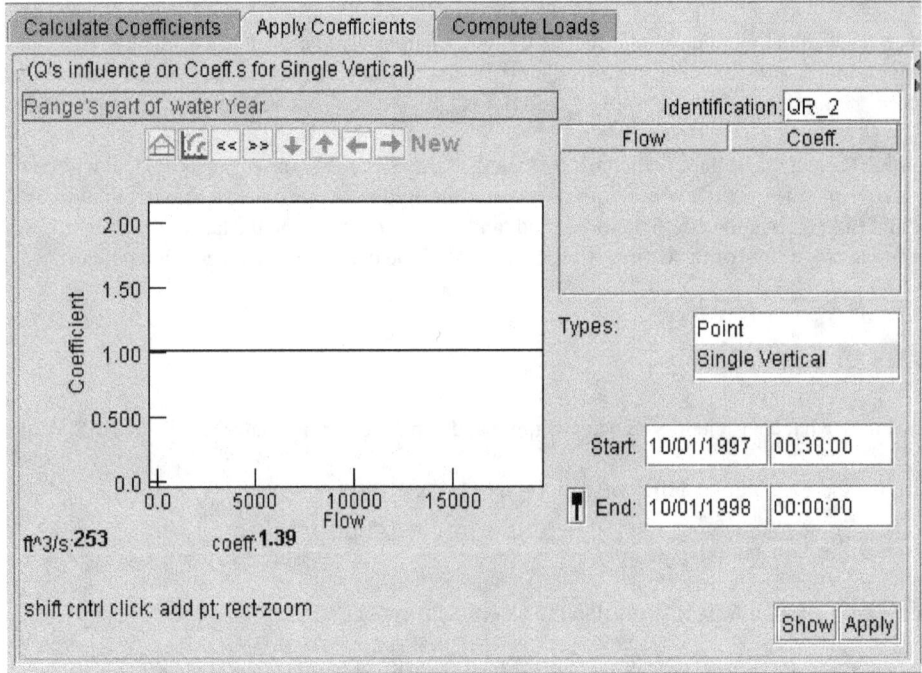

3. The coefficient defaults to 1.00 for the entire year for both point and single-vertical samples (provided that this is the first time you've worked on this concentration file).

4. To modify the default relation by adding coefficients that you have calculated, left-click the coefficient entry in the coefficient-relations panel (causing it to be highlighted; see example below), then left-click the **Add** button to add the entry into the graph and flow versus coefficient table shown in the left panel.

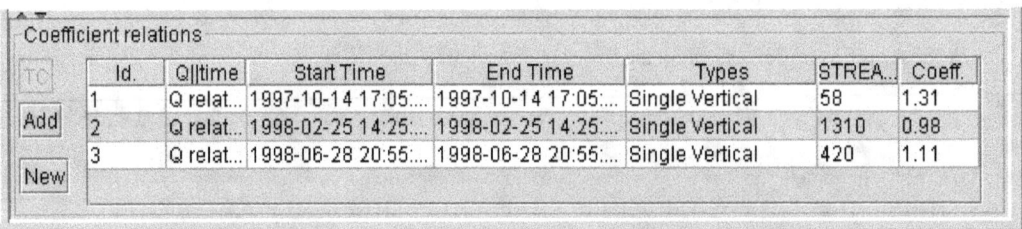

5. To further modify the relation with coefficients that are not in the coefficient-relations panel, enter points directly on the graph by holding the shift and control keys and left-clicking the mouse. (This is exactly the same method used to add estimated concentrations in the working-graph panel.) You can adjust the coefficient in the same way as a concentration data point in the working graph, by either dragging with the mouse, removing the point, or editing the tabular representation to the right of the coefficient versus flow graph (left-click the field that you wish to modify). Figure 13 shows an example of the graph display after adding several coefficients. Note that, for the example shown in figure 13, a coefficient of 1.00 is applied at streamflows less than or equal to 140 ft^3/s, and a coefficient

of 1.54 is applied at streamflows greater than or equal to 4,380 ft^3/s.

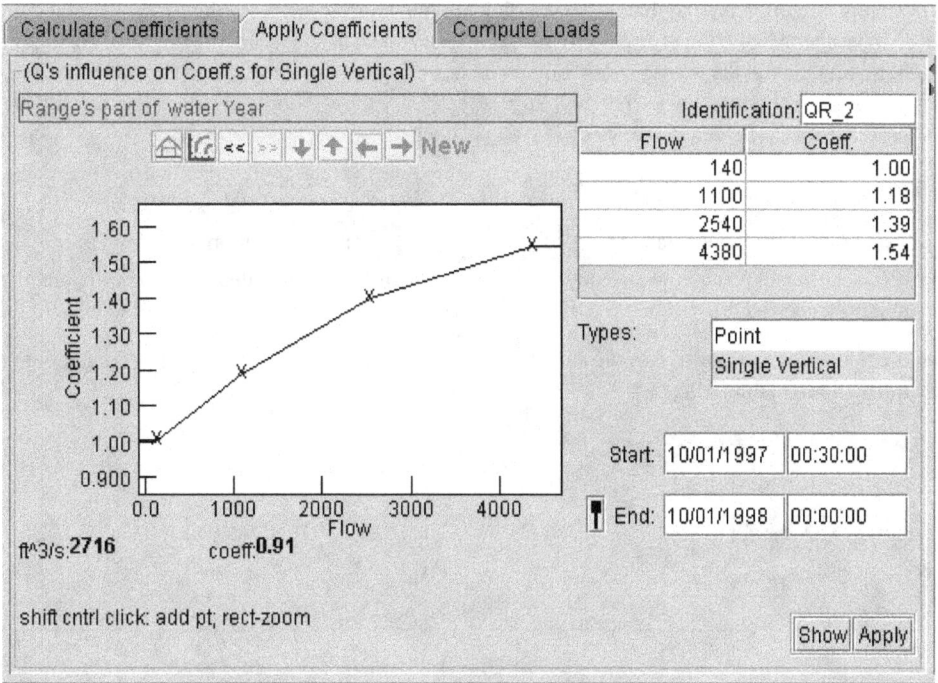

Figure 13. Panel showing cross-section coefficient as a function of streamflow.

6. To preview how the new coefficient relation will be applied, (a) make sure that curve for the point or single-vertical coefficient curves is set to be shown in the working-graph panel and (b) left-click the **Show** button in either the **Apply Coefficients** tab or the Application of coefficient relations panel. The coefficient curve in the working-graph panel (fig. 14) will change to reflect the constructed coefficient relation; however, the concentration curve will remain unchanged.

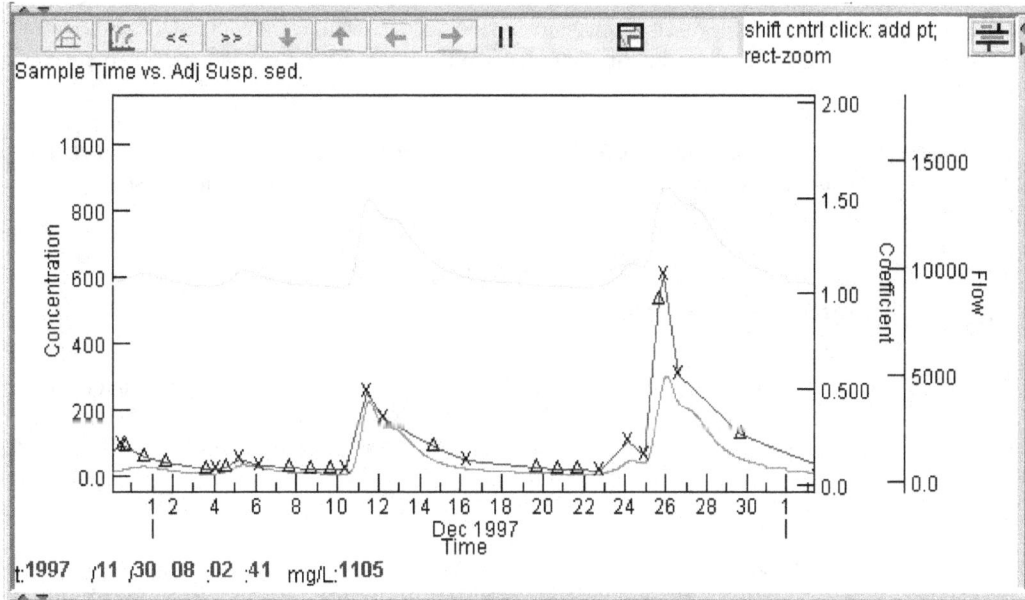

Figure 14. Working-graph panel showing streamflow (red), concentration (dark blue), and cross-section coefficient (light blue) time series.

7. If the coefficient curve is acceptable and you wish to apply the coefficient relation to the concentration data, left-click **Apply** in the *Q's influence on Coeff.s panel*. If you want coefficients applied to estimated points, you must change their representation type to that of the coefficient type being applied. By default, estimated points are assumed to have a representation type of "cross section," and so coefficients are not applied. The representation type is assigned at the time that an estimated value is added. The default representation type can be changed for the duration of a GCLAS session; however, that change affects only those estimated values that are added during the GCLAS session after the default representation type is changed.

8. After a coefficient relation has been applied, make sure that the panel labeled Application of coefficient relations (fig. 15) displays the coefficient-relation icon ▬▬▬ and information describing the applied relation. Note that the identifier shown below the coefficient-relation icon is the same identifier shown in the identification field on the **Apply coefficients** tab (fig.13). If you don't see the entry, it may be necessary to expand the panel by dragging the bottom border down. Use the slider tool to expand or condense the rows containing coefficient-relation information to your liking.

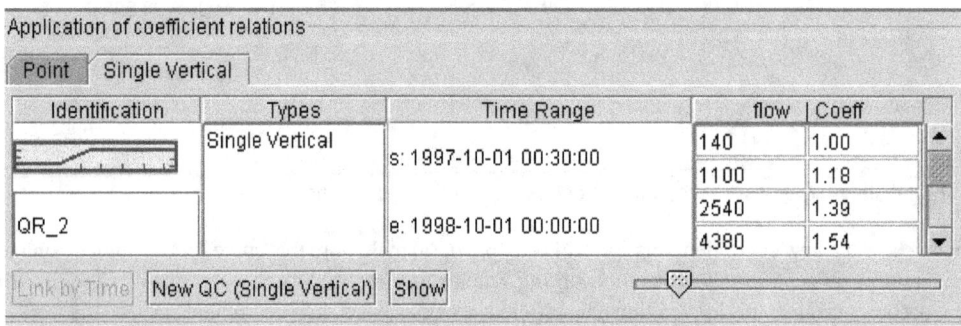

Figure 15. Application of coefficient relations panel showing a coefficient relation that varies as a function of streamflow.

9. If you applied coefficients to single-vertical concentrations and now wish to work with point samples (or vice versa), follow the preceding procedure, but first left-click the **Point** tab in the Application of coefficient relations panel.

10. To create a new relation beginning with a coefficient entry in the coefficient-relations panel, left-click the coefficient entry and then left-click the **New** button to add the entry into the graph and table shown in the left panel; then follow steps 5 through 8 above.

Applying Coefficients as a Function of Time

When coefficients are applied as a function of time, GCLAS determines coefficients from relations specified at a beginning and ending time and then interpolates linearly (as a function of time) to determine coefficients at points in between. The following three steps are the same as for applying a coefficient as a function of flow. The steps after that differ somewhat because you create separate coefficient relations that vary with time, each of which applies over all ranges of flow.

1. Left-click the **Apply Coefficients** tab in the lower left panel after you've moved one or more coefficients down into the coefficient-relations area. The coefficients that were moved in the coefficient-relations area will be listed in the coefficient-relations panel.

2. Left-click on either the **Point** or the **Single Vertical** tab in the Application of coefficient relations panel, depending on which type of coefficient relation you wish to work with.

3. A default coefficient of 1.00 is used for both point and single-vertical samples (provided that this is the first time you've worked on this concentration file). If the coefficient curve has been set to be shown, a straight-line curve representing the default coefficient of 1.00 will display in the working-graph panel.

Selecting and Applying a Previously Calculated Coefficient

4. In the coefficient-relations panel, left-click the row of data that contains the desired coefficient, then left-click the **New** button in that panel. The relation should appear (as a straight line equal to the selected coefficient) on the graph to the left.

5. Use the **Start:** and **End:** boxes (see fig. 13) to set the time range of the coefficient. (See "Appendix 3 — Date and Time Editing Functions" to learn about special functions for editing dates and times.) (Note: To set an approximate time range that you can later modify, left-click the **range** button ▐▐ in the working-graph panel and then left-click in the working graph, dragging the cursor from the desired start time to the desired end time.) To apply the coefficient at a single point in time, you can set the beginning date and time that you want and then left-click on the **take start** button ▍ to make the ending date and time the same as the beginning date and time.

6. Left-click **Show** to see how the changes made will affect the coefficient curve displayed in the working graph and the overview graph, then left-click **Apply** to apply the coefficient relation to the data.

7. Once you've applied the coefficient for the chosen time (or range of time), an entry for that coefficient and time will appear in the Application of coefficient relations panel.

8. Repeat the procedure with any additional previously calculated coefficients.

Creating and Applying a New Ad Hoc Coefficient Relation

9. Left-click the **New QC** button in the Application of coefficient relations panel. A new horizontal curve, default coefficient of 1.00, will display in the Q's influence on Coeff.s panel on the left, with a time range equal to the GCLAS year being worked.

10. To change the coefficient from the default of 1.00, hold the shift and control keys and left-click the mouse to enter a new point. You can adjust the coefficient by either dragging with the mouse, removing the point, or editing the tabular representation to the right of the coefficient-versus-streamflow graph (left-click the coefficient field); you can ignore the entry for flow because, unless you change the curve from a horizontal line, the coefficient will apply over the full range of flows.

11. Left-click **Show** to see how the change you made will affect the coefficient curve displayed in the working graph and the overview graph, then left-click **Apply** to apply the coefficient relation to the data.

12. Once you've applied the coefficient for the chosen time (or time range), an entry for that coefficient and time(s) will appear in the Application of coefficient relations panel.

13. Repeat steps 9–12 for any additional new coefficient relations.

Linking Coefficient Relations by Time

14. In the Application of coefficient relations panel, left-click the icon for the first of two coefficient-time relations you wish to link, then left-click the second one while holding the control key. The **Link by Time** button should become active as shown in the example below.

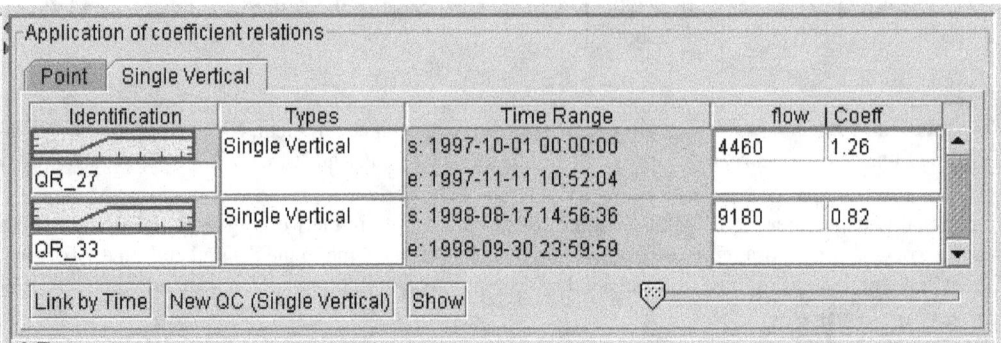

15. Left-click the **Link by Time** button. Within the Application of coefficient relations panel, a new icon indicating the linkage ▱▱▱▱T→▱ will appear between the two linked coefficient relations, as shown below.

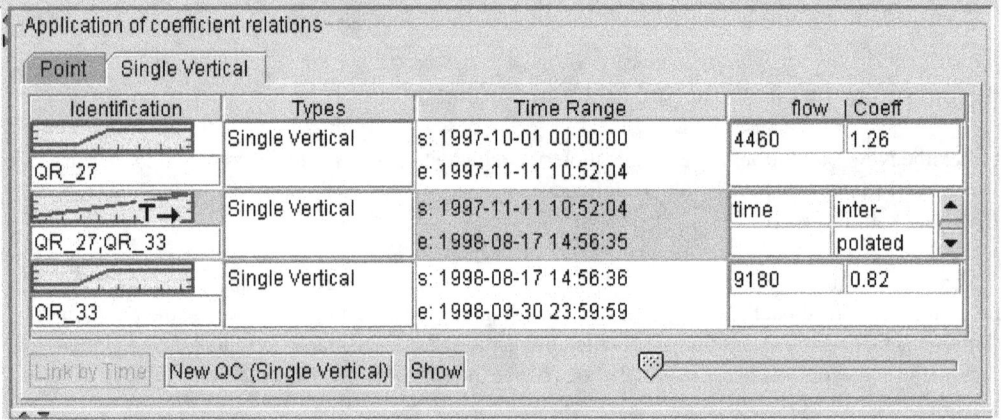

As illustrated below, the panel to the left of the Application of coefficient relations panel will change, displaying (a) a graph on top showing the coefficient-relation as a function of time over the linked period and (b) two graphs on the bottom showing the coefficients relations at the beginning (left) and end (right) of linked period. (The slider tool

▱ allows you to change the flow for which the coefficient-versus-time relation is displayed. The graphs represent the bounding and time-varying coefficients applicable for the flow indicated in the box below the label **"Q:"**. Because, in this example, the bounding-coefficient relations are both applied as a constant over the full range of flows, moving the slider will not affect the top graph but will cause the vertical lines on the bottom graphs to move left or right as the flow value of application is decreased or increased, respectively.)

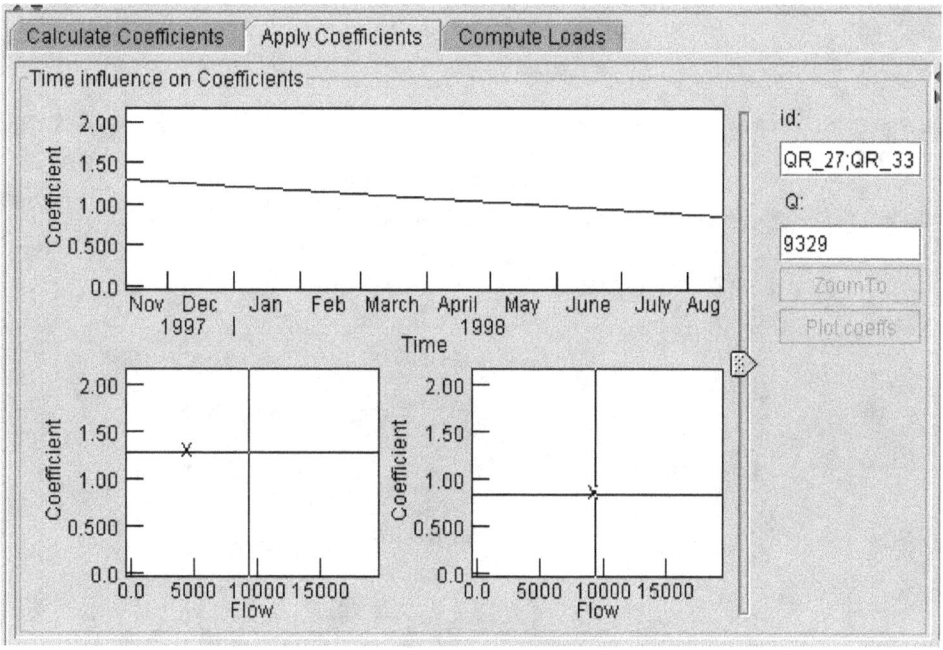

16. Repeat steps 14 and 15 for any other pairs of coefficient relations that you wish to link.

Applying Coefficients as a Function of Streamflow and Time

Constructing coefficient relations that vary with streamflow and time is just a special case of the techniques described in "Applying Coefficients as a Function of Time" (p. 34). The only difference is that the coefficient relations that vary as a function of time are not constant but instead vary as a function of streamflow. For any given point in time, GCLAS determines its corresponding streamflow (by interpolation, if necessary) from the streamflow time series and then computes coefficients for that streamflow from both streamflow-varying coefficient relations that bracket the point in time. GCLAS then determines the coefficient that should be applied at that time by linear interpolation (as a function of time) between the coefficients determined from the bracketing relations. Familiarize yourself with the techniques described in "Applying Coefficients as a Function of Streamflow" and "Applying Coefficients as a Function of Time," then combine these techniques as necessary to get the desired effect on your data.

Computing Loads

At any time during the record-working process, you can compute daily, monthly, and annual loads in GCLAS, literally at the press of a button. If desired, you can easily recompute to test the effects of alternative analysis strategies on computed loads. Load computations are performed by application of the mid-interval method (Porterfield, 1972) to concentration data interpolated (if necessary) to the same temporal frequency as the streamflow time series. Although concentration data do not have to be collected as frequently as streamflow, it is assumed implicitly that the constituent-concentration time series used by GCLAS adequately represents the true time-varying concentration. Generally, this requires that concentration data (or an appropriate surrogate from which concentration can be estimated) are collected relatively frequently.

GCLAS has the potential to compute loads for any constituent that is transported in water at measurable concentrations. At present, GCLAS contains no provision to explicitly handle censored ("less than" or "greater than") data in the load-computation algorithm; consequently, it should not be used to compute loads of constituents that have an appreciable amount of censored values. GCLAS assigns censored values their censoring level when plotting or using the values to compute loads.

To compute loads:

- Make sure that you have added all the estimated points and applied all the cross-section coefficient relations that you wish at this point.

- Left-click the **Compute Loads** tab in the lower panel of the main GCLAS window.

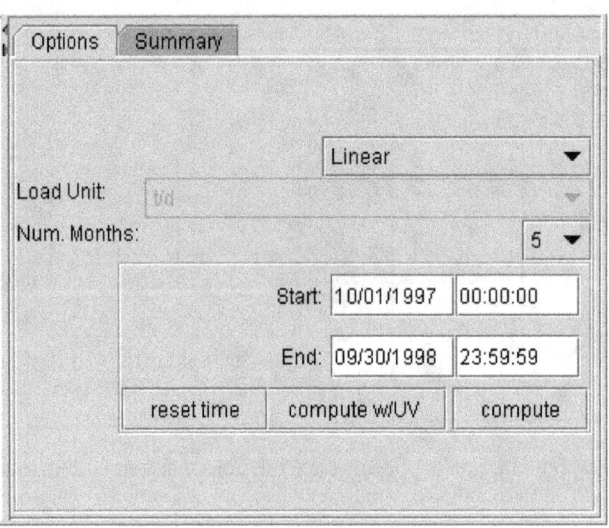

Figure 16. Options tab on the load-computation panel.

- Left-click the **Options** tab in the lower right panel (fig. 16). The following pick-list options are available:

  - Interpolation mode (linear is the default and is recommended; log-transformed linear is provided only for the sake of backward compatibility with SEDCALC files (Koltun and others, 1994))

  - Number of months of data displayed in the loads table (without having to scroll)

- The timespan for load computation is specified by entering date and time values in the **Start** and **End** boxes (**reset time** changes the timespan to the entire water year).

- Left-click **compute** if you're satisfied with the options. Daily load data will be displayed in the left panel. (Note: If the timespan for load computation includes partial days at the beginning or end of the time period, mean concentrations and loads for those partial days will be computed and displayed in the loads table on the **Compute Loads** tab. The partial-day values will also be included in monthly and annual summary values.) If **compute w/UV** is selected, in addition to displaying loads in the left panel, unit values of concentration and load will be written to files in the SEDATA card-image format. (See "Appendix 1 — SEDATA 2- and B-Card-Image Formats" for a description of unit-values card-image format.) Selecting the **compute w/UV** button causes two windows to pop up in which the file names and locations of the concentration and load card-image output files must be specified.

Month	Q	mean C	load
Oct	2060	13	67
Nov	4200	18	343
Dec	31500	89	15900
Jan	94800	126	93000
Feb	70200	112	75000
Mar	69500	126	43300
Apr	62900	155	47200
May	31600	117	24600
Jun	37800	175	35500
Jul	23100	85	8760
Aug	42900	74	22700
Sep	3540	12	142
Ann	474000	92	366000

Figure 17. Summary tab on the load-computation panel.

- Left-click the **Summary** tab (fig. 17) in the right-hand panel to see a monthly summary of the streamflow (in cfs-days), mean concentration (in the concentration parameter's base units), and computed loads (in the load units selected when creating the GCLAS year). In addition, you can compute loads for other time ranges by left-clicking the following buttons:

  to compute loads for the entire water year.

  to compute loads for the current view in the working graph.

  to compute loads for time range specified on the **Options** tab.

Saving Your Work

The **Save** feature allows you to save an in-progress GCLAS session so that it can be resumed later. GCLAS does not automatically save sessions, so use the **Save** feature periodically as insurance against the consequences of a power failure or computer malfunction.

To save a GCLAS session that is in progress:

- Left-click **File** in the menu bar in the upper left corner of the main window, then left-click the **Save** option in the menu that pops up.

- Enter a valid file name in the **File name** field. The extension ".gpf" (where gpf stands for GCLAS project file) automatically will be appended when the session is saved.

- Left-click **Save** to save the file.

- Repeat this procedure as often as desired to overwrite previous saves.

To save a GCLAS session in a different project file (if, for example, you're experimenting with multiple analytical options), use the **Save as** command in the **File** menu (or use the keyboard shortcut **shift-a**).

Note: To reopen a GCLAS project file or open a different GCLAS project file, you'll need to exit GCLAS and restart it.

Exporting Data and Tables

Information can be exported from GCLAS in several forms, including GCLAS gcl files (delimited text files that contain data from the water-quality table), card-image files of computed concentrations or loads, and tables of computed values. Some of the export formats can also be read back into GCLAS by means of the **Import** command on the **File** menu (see page 12).

Exporting GCLAS gcl Files

- Start at the tabular-data panel. If you wish to export only some of the columns in the tabular data, make sure that only those columns are visible.

- Right-click anywhere in the body of the table. A small menu will pop up with **Reports** being one of the menu options. left-click **Reports,** and a rdb report options window will pop up. The options on both tabs of the window are shown in figure 18.

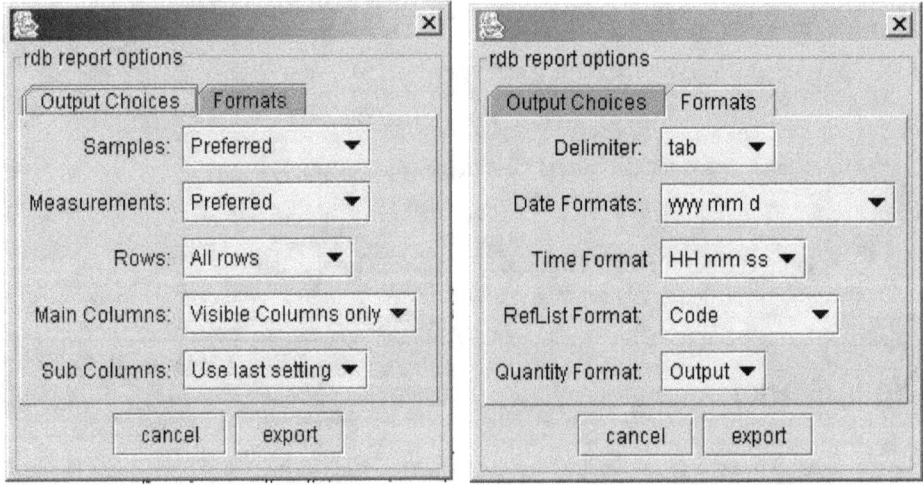

Figure 18. Tab views of the rdb report options window.

- Select the desired options from the pick lists grouped under the **Output Choices** and **Formats** tabs in the dialog box. Generally, you should use the defaults that are displayed, but you may want to change the following:

- **Rows** — choose either all or selected rows

- **Delimiter** — choose tab, comma, or an exclamation point to delimit columns

- **Date** — choose from the following formats: yyyy mm d, yyyy/mm/d, mm/d/y hh:mm:ss, or julian

- **Time** — if the date format chosen above does contain a time field, choose hh mm ss or hh/mm/ss

- Left-click on **export**. The *Export file* window will appear; use it to name and save your exported file. The extension ".gcl" will be appended to the file name specified.

Exporting Concentrations and (or) Loads in Card-Image Format

- Start at the computed loads table in the bottom left panel. Right-click anywhere in the body of the table to bring up the **Reports** command box (fig. 19), select **cardExports**, and then select the desired option to the right.

	Calculate Coefficients		Apply Coefficients		Compute Loads				
Day	**October**			**November**			**December**		
	Q	Conc.	Load	Q	Conc.	Load	Q	Conc.	Load
1	--	--	--	95	13	3.3	514	43	60
2	66	23	0.92	91	14	3.3	380	31	32
3	64	22	1.1	94	21	4.0	204	21	17
4	6	Reports ▸	cardExports	▸	Daily Loads (rounded)				
5	59	18	Print Daily Load Report		Daily Loads (unRounded)				
6	55	16	2.4	72	14	Daily Mean Concentration (rounded)			
7	53	14	2.0	71	11	Daily Mean Concentration (unRounded)			
8	52	12	1.7	68	12	2.1	349	21	19
9	50	11	1.5	68	15	2.7	303	19	15

Figure 19. Load table showing card-image export options.

The Export file window will appear; use it to name and save your card-image file. (Note: Daily loads and daily mean concentrations for days where the specified computation period or available data represent less than 23 hours will not be written to card images even though they appear in the GCLAS loading table.) A description of the daily-values card image format can be found in "Appendix 4 — Unit- and Daily-Values Card-Image Formats."

Exporting Streamflow, Concentration and Load Data in Table Format

Start at the computed loads table in the bottom left panel. Right-click anywhere in the body of the table to bring up the **Reports** command box (fig. 20), and select **Print Daily Load Report**.

	Calculate Coefficients			Apply Coefficients			Compute Loads		
Day	October			November			December		
	Q	Conc.	Load	Q	Conc.	Load	Q	Conc.	Load
1	--	--	--	95	13	3.3	514	43	60
2	66	23	0.92	91	14	3.3	380	31	32
3	64	23					304	21	17
4	64	21	Reports ▸ cardExports ▸				406	23	27
5	59	18	2.9 Print Daily Load Report				689	42	78
6	55	16	2.4	72	14	2.8	535	27	39
7	53	14	2.0	71	11	2.0	405	24	27
8	52	12	1.7	68	12	2.1	349	21	19

Figure 20. Load table showing daily-load report option.

- The Export file window will appear; use it to name and save your daily load report. The daily load report is a text file that can be viewed or printed with any application suitable for viewing and (or) printing ASCII text (for example, Wordpad or Emacs). A small part of a daily load report is shown in figure 21.

DAY	MEAN DISCHARGE (ft^3/s)	CONCEN-TRATION (mg/L)	LOAD (t/d)	MEAN DISCHARGE (ft^3/s)	CONCEN-TRATION (mg/L)	LOAD (t/d)
		October			November	
1	4.7	3	0.02	5.8	3	0.05
2	4.0	3	0.03	5.9	3	0.05
3	3.4	3	0.03	6.7	3	0.05
4	6.5	3	0.06	6.0	3	0.05

Figure 21. Sample of daily-load report output.

Acknowledgments

The authors thank Tom McKallip, a former USGS employee who developed much of the code for GCLAS. We also thank Barry Hill (USGS, Hawaii), Denis O'Halloran (USGS, California), and Michael Roark (USGS, New Mexico), who provided technical reviews and suggestions to improve this report.

References Cited

Edwards, T.K., and Glysson, D.G., 1999, Field methods for measurement of fluvial sediment: U.S. Geological Survey Techniques of Water-Resources Investigations, book 3, chap. C2, 80 p.

Koltun, G.F., Gray, J.R., and McElhone, T.J., 1994, User's manual for SEDCALC, a computer program for computation of suspended-sediment discharge: U.S. Geological Survey Open-File Report 94–459, 46 p.

Porterfield, George, 1972. Computation of fluvial-sediment discharge. U.S. Geological Survey Techniques of Water-Resources Investigations, book 3, chap. C3, 66 p.

U.S. Geological Survey, 2003, User's manual for the National Water Information System of the U. S. Geological Survey—Automated Data Processing System (ADAPS) U.S. Geological Survey Open-File Report 03–123, version 4.3, 407 p.

Appendix 1 — SEDATA 2- and B-Card-Image Formats

SEDATA 2-card format

Column	Description
1	Enter a 2. The 2-card must preceded the B-cards.
2 - 16	Station identification number (left justified)
17 - 28	Blank
29 - 33	Parameter code.
34 - 38	Statistic code.
39 - 80	Blank.

SEDATA B-card format

Column	Description
1	Enter a B. There will be one B-card for each unit value.
2 - 16	Station identification number (left justified).
17 - 20	Calendar year. Four-digit number representing the calendar year in which the observation was made.
21 - 22	Month. Two-digit number representing the month on which the observation was made.
23 - 24	Day. Two-digit number representing the day on which the observation was made.
25 - 30	Time of the observation coded in columns 39-44. Columns 25-26, hour (24-hour system) 27-28, minute 29-30, second
31 - 35	Enter 1440.
36	Remark code.
37 - 38	Blank.
39 - 44	Unit value for date and time listed in columns 17-30.

Appendix 2 — Keywords, Format Codes, and Value Domains for GCL Files

Table 2-1. Time keywords and permissible format codes.

[HH, hour; mm, minute; ss, second; MM, month; d or dd, day; yyyy, year]

Keyword	Permissible format(s)
SampTime	yyyy/MM/d MM/d/yyyy HH:mm:ss
SampTime_Year	yyyy
SampTime_Month	MM
SampTime_Day	dd
SampTime_Hour	HH
SampTime_Min	mm
SampTime_Sec	ss
SampTime_Time	HH:mm

Table 2-2. Sample metadata keywords, format codes, and value domains.

Keyword	Format code		
	name	code	id
SampRepresentation	Cross Section	xSec	I
	Point	pt.	P
	Single vertical	sVrt	S
	Unspecified	unsp	U
SampHydroCond	Not determined	notDet	A
	Stable low stage	lStage	4
	Falling stage	fStage	5
	Stable high stage	hStage	6
	Peak stage	pStage	7
	Rising stage	rStage	8
	Stable, normal stage	nStage	9
SampCollector	Hydrologist	hydr	H
	Observer	obsr	O
	Automatic	auto	A
	Unspecified	unsp	U

Table 2-2. Sample metadata keywords, format codes, and value domains.—Continued

Keyword	Format code		
	name	code	id
SampHydroEvt	Spring breakup	sprBr	A
	Under ice cover	iceCov	B
	Glacial lake outbreak	lakeBr	C
	Mudflow	mudflw	D
	Tidal action	tidal	E
	Dambreak	damBr	H
	Storm	strm	J
	Drought	drght	1
	Spill	spill	2
	Regulated flow	rgltd	3
	Snowmelt	snMlt	4
	Earthquake	quake	5
	Hurricane	hurr	6
	Flood	fld	7
	Volcanic action	vlAct	8
	Routine Sample	rSamp	9
SampMethod	Equal discharge increment	EDI	Q
	Equal width increment	EWI	W
	Grab	Grab	G
	Pumped	Pump	P
	Single Vertical	sVrt	S
	None	NA	N
	Unspecified	Unsp	U
SampQAType	Other QA	other	B
	Spike solution	sSol	8
	Spike	1	1
	Blank	blnk	2
	Reference	ref	3
	Blind	blnd	4
	Duplicate	dup	5
	Replicate	repl	7
	Composite (time)	Composite (time)	H
	Reference material	Reference Material	6
	Not determined	noDet	A

Table 2-3. Parameter keywords, format codes, and value domains.

[XXXXX represents the five-digit National Water Information System parameter code]

Keyword	Format code				
	val	name	code	id	bol
pXXXXXValue	real number[1]				
pXXXXXCoeff	real number				
pXXXXXAdjusted	real number				
pXXXXXStatus_Code		Awaiting review	await	I	
		Presumed Ok	Ok	S	
		Accepted	Accpt	R	
		Rejected	rjct	Q	
		Not reported	notR	A	
pXXXXXType_Code		field	Field	F	
		Lab	Lab	L	
pXXXXXQAType_Code		Not determined	noDet	A	
		Other QA	other	B	
		Composite (time)	tComp	H	
		Spike	1	1	
		Blank	blnk	2	
		Reference	ref	3	
		Blind	blnd	4	
		Duplicate	dup	5	
		Reference Material	refMat	6	
		Replicate	repl	7	
		Spike solution	sSol	8	
		Regular	reg	9	
pXXXXXUsable					TRUE
					FALSE

[1]Parameter values may be preceded by an E, <, or > to indicate that it is an estimate, the true value is less than the indicated value, or the true value is greater than the indicated value, respectively. Values preceded by a < or > are treated in GCLAS identically to uncoded values.

Appendix 3 — Date and Time Editing Functions

Table 3-1. Date-editing functions.

Key	Date Function
Home key	Positions the cursor in the first position of the field.
End key	Positions the cursor in the last position of the field.
Space	There is no effect of pressing this key in the month and day fields. On pressing this key in the year field, it is changed to the default current year. Date-field separator characters are not deleted on pressing this key.
Backspace	On pressing this key in the month and day fields, it changes to some valid month and day, respectively. On pressing this key in the year field, it is changed to the current year. Date-field separator characters are not deleted on pressing this key.
Delete	On pressing this key in the month and day fields, it changes to some valid month and day, respectively. On pressing this key in the year field, it is changed to the current year. Date-field separator characters are not deleted on pressing this key.
Left-Arrow key	Moves the cursor to the left by one character, skipping separators.
Right-Arrow key	Moves the cursor to the right by one character, skipping separators.
Up-Arrow Key	Increments the part (day, month, or year) on which the cursor is positioned. The increment is cyclic, and it updates its super field. For example, if the mask is **dd/mm/yyyy**, the cursor is positioned on **mm,** and the month displayed is 12 and you press the **Up-arrow** key, then the month changes to 01 and the year is incremented by 1.
Down-Arrow Key	Decrements the part (day, month or year) on which the cursor is positioned. The decrement is cyclic, and it updates its super field. For example, if the mask is **dd/mm/yyyy**, the cursor is positioned on **mm,** and the month displayed is 01 and you press the **Down-arrow** key, then the month changes to 12 and the year is decremented by 1.
Character Key [A-Z, a-z]	Does nothing. Causes bell to ring on computer indicating invalid entry.
Digit Keys [0-9]	Displays the digit at its cursor position, and the cursor position is incremented by 1, skipping over separators where they occur. Each field is checked for validity as it is typed. If the field is not valid, then the digit is ignored. For example, if the day field contains the value "11" and the cursor is positioned in front of the first 1, then typing an 8 results in the day field changing to "08" because "08" is the only valid day-field entry that yields a valid result. If a 2 were typed instead of an 8, then the first digit in the day field would be replaced by a 2 (because "21" is a valid day in any month), and the cursor is repositioned following it.

Table 3-2. Time-editing functions.

Key	Time functions
Home key	Positions the cursor in the first position of the field.
End key	Positions the cursor in the last position of the field.
Delete key	**Text selection** — All the selected characters are deleted except for the separators. **No selection** — The character at the current cursor position is deleted.
Backspace key	**Text selection** — All the selected characters are deleted except for the separators. **No selection** — The character before the current cursor position is deleted.
Space Key	**Text selection** — All the selected characters are deleted except the separators. **No selection** — Part (hour, minute, or second) of the time at which the cursor is positioned is deleted. If the cursor is positioned on the separator, then the part preceding it is cleared.
Left-Arrow key	Moves the cursor to the left by one character, skipping separators.
Right-Arrow key	Moves the cursor to the right by one character, skipping separators.
Up-Arrow Key	Increments the part (hour, minute, or second) on which the cursor is positioned. The increment is cyclic, and it updates its super field. For example, if the mask is **hh:mm:ss**, the cursor is positioned on **mm,** and the minute displayed is 59 and you press the **Up-arrow** key, then the minute changes to 00, and the hour is incremented by 1.
Down-Arrow Key	Decrements the part (hour, minute or second) on which the cursor is positioned. The decrement is cyclic and it updates its super field. for example, if the mask is **hh:mm:ss**, the cursor is positioned on **mm,** and the minute displayed is 00 and you press the **Down-arrow** key then the minute changes to 59, and the hour is decremented by 1.
Character Key [A-Z, a-z]	Does nothing.
Digit Keys [0-9]	**Text selection** —All the selected characters are deleted except the separators, the digit at the current cursor position is displayed, and the cursor position is incremented by 1, skipping over separators where they occur. **No selection** — Displays the digit at the current cursor position, and the cursor position is incremented by 1, skipping over separators where they occur. Each field is checked for validity as it is typed. If the field is not valid, then the digit is ignored. For example, if the hour field contains the value "15" and the cursor is positioned in front of the 1, then typing a 2 results in an invalid field because the hour field is interpreted as "25" (and so the 2 is ignored). In this case, if you wish to replace the "15" with an hour value of (for example) "22," then select the entire hour field (so that it is highlighted) before typing.
Separator key	When the user presses the separator key (":") and the cursor is not in the first column of a field, then the cursor is moved to the position following the separator. It the cursor is in the first column of a field, then the separator key does nothing.

Appendix 4 — Unit- and Daily-Values Card-Image Formats

Unit-values (2-card) card-image format

Column	Description
1	Enter a 2. The 2-card must preceded the B- or 3-cards.
2 - 16	Station identification number (left justified).
17 - 28	Blank.
29 - 33	Parameter code.
34 - 38	Statistic code.
39 - 80	Blank.

Unit-values (B-card) card-image format

Column	Description
1	Enter a B. There will be one B-card for each set of up to 6 unit values.
2 - 16	Station identification number (left justified).
17 - 20	Calendar year. Four-digit number representing the calendar year in which the observation was made.
21 - 22	Month. Two-digit number representing the month on which the observation was made.
23 - 24	Day. Two-digit number representing the day on which the observation was made.
25 - 30	Time of the observation coded in columns 39-45. Columns 25-26, hour (24-hour system) 27-28, minute 29-30, second
31 - 35	Readings per day. Right justify, and number evenly divisible into 2880. If the number of readings per day for streamflow is 1440 (corresponding to a 1-minute time interval), then a "Q" must be entered in column 31 of the first (and only the first) B-card image in the data set.
36 - 38	Blank.
39 - 80	Unit value for six consecutive intervals of time. Each value is in a seven-column field (39-45, 46-52, and so on).

Daily-values (3-Card) card-image format

Column	Description
1	Enter a 3. There will be one 3-card for each set of up to 8 daily values.
2 - 16	Station identification number.
17 - 20	Calendar year. Four-digit number representing actual calendar year of data on the card.
21 - 22	Month. Two-digit number representing the month. For example, March is punched as 03, November as 11.
23 - 24	Card number. A two-digit number representing the fraction of the month that the data were collected. This number is coded as follows:

Card no.	Days
01	1 - 8
02	9 - 16
03	17 - 24
04	25 - 31

Column	Description
25 - 80	Daily values. Eight seven-column fields in which the daily values are entered for the designated days. The value should include significant digits to the right of the decimal point where needed. Whole numbers need no decimal point. Blank fields are interpreted as "no data available."